The New Energy Movement

The New Energy Movement

How Revolutionary New Energy Inventions can Transform our world

JEANE MANNING
JOEL GARBON

Viva Books
New Delhi | Mumbai | Chennai | Kolkata
Bangalore | Hyderabad | Kochi | Guwahati

First Indian Edition 2010

VIVA BOOKS PRIVATE LIMITED

- 4737/23 Ansari Road, Daryaganj, New Delhi 110 002
  E-mail: vivadelhi@vivagroupindia.net, Tel. 42242200
- 76, Service Industries, Shirvane, Sector 1, Nerul, Navi Mumbai 400 706
  E-mail: vivamumbai@vivagroupindia.net, Tel. 27721273, 27721274
- Jamals Fazal Chambers, 26 Greams Road, Chennai 600 006
  E-mail: vivachennai@vivagroupindia.net, Tel. 28290304, 28294241
- B-103, Jindal Towers, 21/1A/3 Darga Road, Kolkata 700 017
  E-mail: vivakolkata@vivagroupindia.net, Tel. 22836381, 22816713
- 7, Sovereign Park Apartments, 56-58, K. R. Road, Basavanagudi, Bangalore 560 004
  E-mail: vivabangalore@vivagroupindia.net, Tel. 26607409, 26607410
- 101-102 Mughal Marc Aptt., 3-4-637 to 641, Narayanguda, Hyderabad 500 029
  E-mail: vivahyderabad@vivagroupindia.net, Tel. 27564481, 27564482
- First Floor, Beevi Towers, SRM Road, Kaloor, Kochi 682 018, Kerala
  E-mail: vivakochi@vivagroupindia.net, Tel: 0484-2403055, 2403056
- 232, GNB Road, Beside UCO Bank, Silpukhuri, Guwahati 781 003
  E-mail: vivaguwahati@vivagroupindia.net, Tel: 0361-2666386

www.vivagroupindia.com

Published by arrangement with

Amber Bridge Books
101-1001 W. Broadway, Suite 162
Vancouver, British Columbia V6H 4EH

ISBN: 978-81-309-1481-7

Published by Vinod Vasishtha for Viva Books Private Limited, 4737/23, Ansari Road, Daryaganj, New Delhi - 110 002. Printed and bound by Raj Press, R-3, Inderpuri, New Delhi - 110 012.

To the many generous souls who personally helped the authors continue on our journey of discovery over the recent years and to those who educated us. Even though you are more numerous than we can list here, your names are each written indelibly in our hearts and illuminated with our gratitude.

CONTENTS

Preface ix

Part I Wake-up Time 1

1. Jeane's Journey 3
2. Reclaiming Decision-making 17
3. Breakthroughs Needed 28
4. The World We Can Choose 42

Part II Hidden History Impacts You 53

5. Nikola Tesla 54
6. Surging Sea of Energy 65
7. Living Energy 72
8. Power from the Universe 82
9. Magnetic Pioneers 92
10. Oil and Water 106

Part III Emerging Era 117

11. Breakthrough Theories 119
12 The Water is Hot 134
13. Better Than Fire 148
14. Singing Water 159
15. Magnets in Your Future 169
16. Downpour of Solutions 176

Part IV Who Hijaçked the Energy Revolution? 187

17. Misled Media 189
18. Hitting the Wall 201

Part V You Can Help Reclaim Choice 211

19. Opening Doors 213
20. What You Can Do 226

Appendix 1 Energy Innovation Act 237

Appendix 2 Mohorn's Self-running Wall Dryer 243

Notes 248

Resources 259

Glossary 261

About the Authors 264

Index 266

PREFACE

The quantum-leap clean energy revolution has a much better chance than it had a dozen years ago when Jeane Manning's book *The Coming Energy Revolution* was published; conditions are different now:

- A fast-growing percentage of people are aware of atmospheric pollution from carbon fuels and at the same time are weary of oil wars.
- Many are worried about nuclear waste from power plants and those facilities' vulnerability to terrorism.
- Scientists are concerned about the increasing acidity of oceans as the waters soak up excess atmospheric carbon dioxide.
- Vehicle owners are increasingly alarmed and financially distressed by the escalating price of fuel.
- People see the cost of corn-based food shoot up as a result of the agribusiness race to sell "biofuels" that are based inappropriately on corn.

- The Internet allows the global brain-trust of inventors, engineers, other researchers and activists to make connections at a previously impossible speed.

The push for an energy revolution

The result of that global brain-trust is an international effort to completely change the sources of energy on which our world is based, and to tap into a previously unrecognized background energy found everywhere. Some of its proponents call it "free energy." Some call it "zero point energy." By any name, it has the potential to affect the life of every human being on Earth.

Background energy—another term we will use in this book—is real, despite having been publicly confused with the old discredited idea of perpetual motion. As with the earlier personal computer revolution, inventors are making breakthroughs in home workshops and garages as well as in professionally run laboratories. Observers of these developments predict that this revolution will have more of an impact than PCs have had. These inventions could do more than transform our homes, vehicles, and factories; they could also help clean up the water, air, and soil.

The co-authors of this book have spent more than thirty years combined in learning about energy-technology alternatives and what they mean to society. Joel Garbon's career has been in industrial science, and some of his colleagues hold the view that certain new energy systems would violate a principal law of physics if the inventions work as claimed, and therefore must not be real.

For instance, certain technologies are claimed to produce more measurable output power in the form of electricity, heat, or mechanical power than can be accounted for by the measurable input power. On the surface, such a claim would be contrary to well-established laws of thermodynamics, such as the law of conservation of energy.

However, Garbon and others have investigated such inventions and have learned firsthand that many of these prototype devices do work as claimed. So from a practical standpoint, they obviously must not violate any real *laws* of nature. They may, however, violate some human-conceived *theories* of nature. But the explanation for the unusually high power yields may be even simpler. New detecting and measuring equipment may be needed to understand what is happening in such machines, as we learn in chapter 11.

Tools for change

We're not promoting products; in general the products are not even on the market. At the time of this writing you can't find a line of free-energy converters for sale to replace the home-power generators on the market. No auto salesroom displays an electric automobile powered by the cosmos. No manufacturer advertises an Air-Cleansing Water-Fueled Truck.

That situation can and inevitably will change, because of breakthroughs being made. For instance, a device has been successfully selling in Europe to deal with moisture that creeps up walls inside stone-walled basements. It cannot power household appliances, but does do useful work. The Mohorn device (see Appendix 2) taps into a previously unrecognized source of ubiquitous energy.

This book introduces new energy systems as tools for creating change, not as a cure-all for human problems. Powerful tools require utmost respect. You don't hand an electric drill to a child; you wait until the child matures and is more responsible. However you do educate them along the way to that time when they can handle and use the tool. Similarly, the human family's general level of responsibility to all life is evolving, and perhaps humankind will become more responsible as we face the environmental crises caused by human actions—from actions of overfishing the oceans to degrading the atmosphere with polluting toxins.

Who invents breakthroughs

Outside of secretive government or industrial laboratories, inventors in this field have often been individuals without advanced scientific training, working in small workshops. The standard scientific viewpoint has been that these inventors didn't know what they were doing, that these new energy sources cannot exist because they go against the known laws of physics.

In recent decades, however, some highly trained scientists have defied that viewpoint and started taking new energy seriously. Around the world, respected physicists are recognizing that official science has painted itself into a corner. For too long, orthodoxy ignored mounting evidence in support of new energy. Now it seems as though the laws of physics will have to be interpreted in a new way.

What's in it for non-techies

We're writing this book for people who are not necessarily technically inclined, for the following reasons:

- Society's choices of energy technologies affect everyone.
- Voters need to know that oil wars are unnecessary.
- Youths need hope for a better future.
- The new science can promote world peace by uniting humankind in the knowledge that all peoples are profoundly inter-connected in a sea of energy.
- New Energy books aimed at technical experts do exist but other people want simpler explanations such as those provided here. People deserve to join the debates.

What to do with the knowledge?

If you are an opinion-molder or policy maker, this book can help you draw up a citizens' map of a route to a better future. For instance, enviro-activists may see beyond a "sustainable" to a "restorative" economy in which abundance of clean energy sources allows us to restore ecosystems and to seriously clean the air, water and soils. And with energy literacy, citizen's groups get the full picture of potential energy supplies.

If you are technically inclined, consider the clues pointing toward how to create revolutionary energy systems, and explore the websites on the starter Resources list at page 259 . Teachers can give hope for the future to students, and be motivated to encourage the science whiz to develop his or her inventive talents.

If you are an economist, politician, social scientist, community leader or are involved in national or regional planning, you will be able to help create a new society where "disruptive" (disruptive only to the profits of entrenched industries) inventions are allowed and are used for the betterment of the people and all life.

If you serve elsewhere in society, you can help spread the word—telling your friends, family and associates: "There are better ways to go. We don't have to resign ourselves to a brutal future. We can each contribute to a better world, if only by the power of positive thinking."

Meanwhile, of course, you can also be an example of walking-the-talk environmentally. Many excellent books tell you how to reduce your ecological "footprint".[1]

To picture the changes the new energy Galileos expect to bring about, imagine yourself buying an enhanced energy converter about the size of a laptop computer. This fuel-less device contains no moving parts, yet it puts out enough power to run your home or your new electric vehicle without being plugged into a wall socket or a battery. Since you no longer have to pay a utility bill or buy gasoline, you have the money to lease or purchase the converter. After the hardware is paid for, the electricity you use is free. You can live anywhere, from a mountaintop to a houseboat, because you can heat, air condition and power your home cheaply.

When can you buy a new-energy device? That depends on factors discussed in this book. A lot of new-energy hardware is in the crude pre-manufacturing stage—where the aerospace industry was in 1903 when the Wright brothers flew their homemade aircraft for less than a minute along a beach. However, collaboration and some substantial investment could bring some of these inventions to store shelves soon. China, India, Japan and a few other countries without oil wells—countries strongly motivated to find new sources of energy—show the most interest in such a team effort.

Skeptical

Is harnessing the energy of space an impossible dream and are its proponents merely kooks, as new-energy debunkers would have you believe? Are the guardians of official science correct to say it is impossible to run machines on water, much less on energy from thin air? Such expectations are reinforced when looking at some of the amateur literature of what is called "fringe science". However, as the authors looked more deeply, increasingly it did look as though it was possible to convert a previously unrecognized energy source into useable power. Now the weight of evidence has convinced us of the reality of useable new-energy inventions.

This book is intended to be a discussion-starter. Issues raised by the prospect of cheap electrical power and decentralized sources of abundant, clean energy, and other implications of this fascinating world should be explored and discussed publicly. They are crucial to the economies of countries, and to the well-being of individuals.

Lone inventors

The inventors you will meet in this book represent only a small portion of the new-energy scene. While this book champions the lone inventors and mavericks, we do not mean to underestimate contributions from the worlds of academia, government, and business. These institutions, though, are backed by well-financed public relations efforts. This book is intended to balance the picture.

We relate the stories of these science renegades to not only explain new-energy theories and devices, but to also show the harassment these inventors have encountered. Our aim is not to arouse an "Ain't it awful" reaction. Instead, we wish to draw public attention to the situation, in the hope that public understanding will smooth the path of these energy visionaries. We all have a stake in their success.

Even now, the winds of change are blowing through the smog of our past ignorance. Many brilliant minds around the world are making breakthroughs in revolutionary energy technologies by using a variety of approaches. It's breakthrough time.

PART • I

Wake-up Time

When they hear about the international citizens' network called the New Energy Movement, our acquaintances often respond as if there is no urgency:

> Ah yes, new energy technologies. Green is in! Thanks to Al Gore, everyone is talking about climate change and clean energy. Investors are lining up to fund renewables. The time has come.

The era of renewable energy has *not* arrived. The energy sector growth statistics show that the Green Era won't really be here until the use of nuclear power and fossil fuels shrinks relative to the use of clean energy systems. And the vested interests in unclean energy won't back away from trillion dollar profiteering until the people insist.

Longtime researcher of energy issues Andrew Mount cites statistics showing "we absolutely need an energy revolution to avoid eroding the environment to the point of a scenario like we saw in the film *The Day After Tomorrow* . . . Even if we were to make incremental gains in the renewable-energy sector—of 300 per cent increase in renewable and clean energy technologies between now and 2020—we'd only reduce oil dependence by two per cent."

Part I outlines the urgency of the need for clean energy-abundance breakthroughs, and helps you picture in what ways they would transform your world.

1

Jeane's Journey

We are on the rim of a new era of energy conversion. The opposition is big money and decision makers in industrial conglomerates who stick to everything which today guarantees profits and power. Only big groups of people with awareness will be able to overcome this gigantic fortress.

—GOTTFRIED HILSCHER,[2] *German technical editor*

It just might be that the discovery of the vacuum energy as a limitless energy source is to be synchronized with a spiritual renaissance for all of humanity.

—MORAY KING, *Industrial engineer and New Energy researcher*

Breakthrough Power is written by a woman trained to be a people-watcher and a man schooled in what are called the hard sciences; we are a sociologist and a chemist—appropriate for a topic that affects everyone's future.

Joel Garbon and I met in 2003 while helping establish the New Energy Movement, a grassroots effort promoting clean low-cost local systems for tapping into nature's abundant energy. Although close colleagues in the movement, we are a generation apart in age. While Joel was studying chemistry and engineering and earning his university degree in applied science, I had my degree in sociology and years as a journalist.

This book begins in the first person because the few glances back at my earliest experiences in the "energy underground" are intended to encourage others to make the easy journey from technological indifference to a reasonable level of energy-literacy. However we believe it is even more important to see the big picture—the high stakes for humankind that are involved in its *choice* of technologies. And the high-stakes importance of a widespread change of attitudes.

You don't have to go as far as to attend dozens of science conferences, travel around the world to interview scientists, dedicate hundreds of hours to transcribing audiotapes, or empty your wallet for energy-related books, tapes and DVDs. You don't need an academic degree in hard science nor a career in a technical industry as Joel has. All you need is a passion for cleaning up Earth and creating a better civilization—a sharing, just and peaceful society.

New-Energy journey begins

I first heard about non-conventional energy inventions in 1981 while with friends in a restaurant in the interior of British Columbia, Canada. One of them was a musician who had just returned from the nearby city of Penticton where he had been introduced to an electrician named Bill Muller. His impression of the electrician's work was stunning:

> They call him the 'Magnet Man'. He sold everything he owns to work on his fuel-less electricity-generating invention . . . You've got to hear what you can do with these new magnets . . . very strong permanent magnets. You can use them for fifty years and they don't lose strength . . .

My first reaction was to tune out. Techie talk.

The musician was sketching on a napkin as he enthused. "He found a way to spin magnets past wire coils on a wheel without electrical drag. He invented a configuration which allows easy movement of the wheel without getting a magnet stuck opposite a coil and its core . . . "

Blank faces across the table momentarily silenced the musician.

"Look," he said patiently, "you all dream about living out in the mountains, in pristine meadows way beyond the power lines. But you still want electricity to run your stereos. How will you get reliable power? Solar won't do it; the sun doesn't shine there half the winter. Wind power isn't reliable in Bear Valley. And you don't want a noisy fuming diesel generator.

"What if there's an invention any mechanic could build? It sits there and makes enough electricity to run itself and have plenty left over for appliances, even electric heaters for a greenhouse. You wouldn't have to chop wood . . . "

One of the men at the table smiled smugly. "An impossible perpetual motion invention . . . "

"No! This guy knows that perpetual motion is impossible in any closed system—you can't get more out than goes in. But he says his machine is an open system. Open to the cosmos. The magnets constantly replenish their strength by tapping into some primordial background field of energy . . . "

Unimpressed, the skeptic stroked his beard. "Go see for yourself," the musician challenged.

A petite woman at the table reported that when she had lived in his town, the inventor had shown up once at her metaphysical study group. She remembered his questions about "prana", a word coined in ancient India meaning an all-pervasive subtle form of energy—the invisible background that permeates every cell of our bodies. Bill Muller had wondered if the prana concept might solve his puzzle: how do permanent magnets perform work –exert pushing and pulling forces indefinitely—without being depleted? Do magnets tap into some source of energy unknown to today's science?

A massage therapist in the group had also met Muller. The therapist's own career convinced him that a beneficial life-force "subtle energy" is real, even though scientists didn't have instruments that could detect it. Alternative medicine views our bodies and cells as an expression of energy which comes from an underlying universal substance or super-fine level of vibration, he mused. Could magnets also interact with a different form of energy than standard electricity?

I started to wonder. *What if this unusual motor thing is real, and could in fact halt destructive megaprojects? The damming of wild rivers for instance. Could there really be an alternative clean power source that's way more effective than solar or wind? If so, then no more oil wars!*

Maybe it's time to learn a bit about electricity.

Magnet man

In the next month several of my male friends and I arranged to meet the Magnet Man at his home. It turned out he was far from having a market-ready power generator, but it was a meeting that changed my life journey.

Bill Muller and 1980s magnetic experiment.

Bill Muller was a tall grey-haired man with an apparently high level of vitality despite incessant cigarette smoking. He moved around the room lightly, with nervous energy.

Holding everyone's attention by his forceful personality, he described his invention as a way to make a heavy wheel carry strong magnets past electricity-inducing copper coils without needing to fight the electrical drag force which usually opposes rotation and limits how efficient a generator can be. His wheel didn't have any "stuck" position; it moved freely.

"We have a magnetically balanced flywheel."

In his basement workshop, Bill showed us the beginnings of a permanent-magnet generator.

"Don't call it 'perpetual motion!' The 'fuel' to power the generator is the strong magnetism of rare-earth magnets," he said, while weaving his way through a clutter of machinery with the ease of an athlete. He quoted nineteenth century science pioneer, Michael Faraday: "'Any change in the magnetic field around a wire tends to set up an electric current in the wire.'"

I couldn't follow his rapid-fire discussions of motors and generators, so he turned his back and addressed the male visitors. They later admitted to having just as much difficulty following his explanations.

I mentally reviewed the little I had learned: *electric current begins running in a wire when the wire is near a changing magnetic field.* In my life so far, electricity had seemed a topic only meant for the pylons-and-hardhats crowd.

Benefits for the people

Parts of the inventor's message were easy to grasp, however.

> Can you imagine what this will bring to mankind? Think what you could do with something that's just powered by magnets! You don't need the power company. You could live anywhere! Grow food all winter . . . When we get the machine self-running you can unplug from the electrical grid. Unplug from the utility bill!

He said mainstream science can't explain what force continually replenishes the field of powerful permanent magnets. Such magnets can lift many times their own weight and they only lose strength if struck forcefully as with a hammer, or if subjected to extreme heat. What gives magnets the strength to do work? The answer is at the atomic level. Then what gives atoms their perpetual spin?

One of my friends interrupted Muller's line of questioning by asking if the motor/generator would be able to run a car.

"Sure, we can make one that'll run a car. We just need some money to develop all this stuff."

We were disappointed that Muller's company Pran Technology was still in the research phase. Afterward one skeptical friend warned me against investing in a venture that could be a bottomless pit sucking up my time and money.

Food and air at risk

Nevertheless, I felt a sense of urgency. Our provincial government was making noises about building another huge hydro dam in the Peace River valley—flooding fertile bottomlands to make electricity. I had spent my early school years on my family's farm, and learned to value loamy alluvial soil that can grow mineral-rich food. *Don't drown farmland under tons of water behind a dam! Living soil is disappearing fast enough*, I thought. And in cities, fresh clean air had faded into a memory, due to the burning of carbon fuels.

"*If Bill is right, we don't need oil or megaproject dams or nuke power plants at all*," I realized.

My impression that the impatient inventor had been pat-on-the-head patronizing toward a woman's intelligence only increased my resolve to find the truth. I knew that society's choice of its power source affects everyone—men, women and children.

Back home, I headed to the public library. How does a non-technical person learn about a technical topic? I started in the children's and junior high school sections on electricity and magnetism, and worked up from there to weightier books.

Look at me, a social worker people-person do-gooder. What's with this new-found technophilia? Despite enjoying a laugh at myself, I felt a shift inside me, as if a doorway had opened to a mysterious and inviting new view.

When my youngest son and I moved to Penticton two years later, my education escalated. I volunteered to type letters for the busy Muller household in return for the opportunity to read their books and meet the visitors who came to Penticton from around the world. *I'm not a scientist and can't judge whether 'free energy' is possible, but as a writer I can tell the world about this fascinating scene,* I thought.

Archetypal inventor's story

Bill Muller's life story has many elements in common with those of other inventors I would later meet. Born Wilhelm Johann Friedrich Muller in 1931 in Bremen, Germany, from early on he felt a strong connection to the natural world and an interest in science, and had dreams of becoming a forester. But with the harsh realities of living in a devastated Germany in the aftermath of World War II, it was wiser to learn a trade. Eventually employment as an electrician at Siemens Corporation took him to South America and to Canada.

His fascination with magnets—as possible channels for diverting an unseen energy source—began in 1968 when he and his wife Ilona lived in Barrie, Ontario. He had salvaged two 30-pound bullhorn magnets from a radar installation and set them on his workbench. Idly dropping marble-sized steel ball bearings into an empty sheath made for holding golf clubs, he then held the transparent tube between the large magnets. Inserting another bearing, he was fascinated to see the steel ball pause in the plastic tunnel and vibrate. Apparently the "steelie" was in the middle of a strong magnetic field or fields. Suddenly the steel ball was launched out of the tube at such velocity that it blasted through a concrete wall and disappeared.

Muller intuited that the ball's acceleration had something to do with the frequencies at which it had vibrated, and he believed he could somehow harness the mysterious power of magnets into a revolutionary and highly useful invention.

A turning point in his research came in 1981, the year I met him, when he bought some powerful samarium/cobalt magnets the size of ice cubes. They were a revolutionary advance in the strength of magnets; their holding force would have overwhelmed and stopped the spinning of a rotor wheel if he had not been inspired to rearrange parts of the machine:

If he had followed the usual pattern and placed the super-strong magnets in his wheel across from an equal number of copper coils, immense force would have been required to wrench the wheel forward. He solved the problem of the stuck position by simply eliminating one of the pieces and distributing the rest of them evenly, so that 16 magnets rotated past only

15 coils/cores. This odd-and-even configuration of coil cores and magnets offset the holding force. Now the wheel spun easily in either direction.

Had he eliminated work previously required in a motor/generator? Muller thought about the implications of doing so; it means that society could tap into the power of magnetism instead of burning polluting fuels to turn a generator. Increasingly he felt he was on the trail toward new clean energy technology that could help rejuvenate the Earth.

What's the point of zero point?

During their years in Penticton, the guest book and correspondence at Mullers' home was an education in itself. Bill Muller showed me a letter from engineer Rudolf G. Zinsser of West Germany[3] who agreed with a Canadian engineering report that said Muller's invention was ideal for getting more electrical power out of windmills.

"By the way," Zinsser's letter advised, "the notion 'efficiency' applies for a closed physical system and cannot by definition exceed 100%. (Instead of saying efficiency) you want to state 'yield', which applies to open physical systems such as a windmill whose energy yield exceeds by far the mechanical work put into turning the windmill's blades into the wind."

"Closed . . . open." The arguments are all about the law of conservation of energy, I realized. *Does the no-free-lunch law of physics cover every situation?*

Zinsser's letter went further: "As for the alleged source of energy, I am convinced that the so-called zero-point energy does exist. There is plenty of scientific literature on the subject by recognized physicists."

'Zero point energy'? What's that? The letter answered my question in part: "Quantum gravity, as it is also named, is 'alive' within spaces as small as 10^{-23} centimetre."[4] Zinsser said such as space is far smaller than the diameter of a nucleus of an atom. The German scientist had worked through the mathematical calculations in his own invention an attempt to explain its surprisingly high yield of mechanical energy.

I was puzzled about zero point energy, but more information would soon come to me when I met a former consultant for the National Aeronautics and Space Administration (NASA).

Rocket scientist

One day the Muller household buzzed with news. "Dr. Schaffranke's coming for a visit!"

Rolf Schaffranke PhD was a distinguished scientist who had retired to the American state of Georgia. Bill Muller was impressed because Dr. Schaffranke had been in Peenemunde, Germany, where he worked alongside the legendary rocket scientist Wernher von Braun. Schaffranke was the youngest of the group of German scientists including von Braun who were spirited out of Germany by American intelligence agents at the close of World War II and brought to the United States in order to advance the fledgling U.S. rocket program. Schaffranke worked for the aerospace industry, and before retirement was on a NASA contract as a consultant to the propulsion laboratories in Huntsville, Alabama.

Before the visitor arrived, Bill pointed to the bookshelf and told me that *Ether Technology* was written by Schaffranke.

"It says the author is 'Rho Sigma'," I noted.

That was a pseudonym, Ilona Muller replied. When Schaffranke wrote the book, he was still working for NASA. Previously his NASA boss had reprimanded him for publishing a historical review of inventions. The boss had written in Schaffranke's personnel file, "It is restated that any journalistic activity along this line should be submitted for approval through company management".

'Expert' predictions

What had been so objectionable about Schaffranke's article in the engineering journal? He had merely pointed out the numerous instances in which science experts have held onto an incorrect out-dated worldview and thereby stifled scientific advancement and breakthroughs. He had looked into history as far back as 16th century physicist Sir William Gilbert, who said "Science has done its utmost to prevent whatever science has done."

The book *Ether Technology* gave many examples of experts rejecting what they didn't yet know, including:

- Von Braun's teacher had written a book about space rockets and sent it to ten publishers, who each sent it back. Schaffranke doubted that they read more than the title before deciding the book didn't fit the accepted worldview. An expert reviewed the book and arrogantly said the time had not yet come for space rockets "and indeed probably never will come".

- American physicist R.H. Goddard was ridiculed and called "Moon-mad Goddard" until he launched the first successful liquid-fueled rocket.

"This history is eye-opening," I said, "but I don't view rocket scientists as the apex of intelligence!"

"You mean you boycott the venerable expression 'It doesn't take a rocket scientist . . . '?"

"It doesn't take a biologist to figure out that launching an oversize burning firecracker isn't the best way to get to the stars."

Laughing, Ilona responded, "Burning fuel? There *is* something better."

Schaffranke's book shattered my illusions about scientists' objectivity. The former NASA consultant quoted a colleague about how uncomfortable it is for "scientists with a hard academic glaze" to meet up with any discovery that disrupts long-established dogma. He said these experts become like the ancient map-makers who wrote on their maps, near the Pillars of Hercules (Gibraltar), "hic deficit orbis". Here the world ends.

The physicist gave another instance. Respected scientists of the early 1930s insisted that any attempt to use the energy contained in an atom's nucleus would be doomed. These authorities proclaimed that the energy released would be less than the energy required to smash the atom.

Schaffranke had looked to *The Truman Memoirs* for another example. Admiral William D. Leahy was Chief of Staff when he told the president of the United States that an atom bomb project was impossible. "The bomb will never go off, and I speak as an expert in explosives." A short time later, Hiroshima and Nagasaki tragically showed how wrong that expert was.

These history lessons made me question further. *Could today's 'experts' be blind to a new clean-energy science? Could academics overlook a beneficial paradigm just because it's so very different from what they've been taught?*

Gentleman physicist

When I met Rolf Schaffranke, he turned out to be a cultured, gracious gentleman and curious about my keen interest in energy.

I told him about a small but eye-opening meeting of the Canadian solar industry I had recently attended as a newspaper reporter. There I saw a blatant contrast. On one hand I knew the public had been led, by outside "experts"—to believe that solar technologies will never be efficient enough to provide cost-effective electricity. On the other hand I heard frustrated workers in the solar industry present a very different story.

Even in the 1980s researchers had advanced the efficiency of solar photovoltaics—electricity from sunlight—far beyond what the public heard about, but solar didn't get enough government support to build solar panel factories that could bring down their prices. Meanwhile their government gave the oil and nuclear power industries the bulk of the energy sector's taxpayer-derived financial support and subsidies.

Schaffranke nodded in apparent sympathy when I summed up my motivation. "I don't want my grandchildren to suffocate in an oxygen-deprived world as a result of fuel-burning, or be sickened by radioactive garbage in the air or in the waters."

The physicist said he believed Bill Muller's invention could be developed into a high efficiency nonpolluting generator of electricity. Its simplicity meant it would be fairly easy for other skilled people to build.

I remarked that an engineering consultant from Alberta said the Muller generator should be put in windmills to make more electricity per gust of wind. Schaffranke agreed. If developed, Bill's invention could dramatically improve the electrical output from windmills.

I asked the distinguished guest what motivated him to venture outside the accepted mindset. "What caused a rocket scientist to point out flaws in established theories about how the universe works?"

He replied that he cared about the fact that important new science is being ignored—new science pointing toward fuel-less generators and vehicles.

Hitting the road

Rolf Schaffranke was the first of dozens of credentialed scientists I've met on the new-energy trail. In 1986 the Muller family invited me to join them on a road trip in a borrowed motorhome—to southern California for a new-energy conference sponsored by a magazine titled "Magnets In Your Future".

Since that first conference I've attended more than two dozen such meetings in a half-dozen countries. I'm grateful to have met so many brilliant inventors and researchers and remarkable visionaries, and encouraged to see younger-generation colleagues such as co-author Joel Garbon—a warm and natural speaker as well as a knowledgeable scientist—providing an increasing presence on the new-energy conference circuit.

The benefit of our travels, to readers of this book, is that we can share what we have learned from inventors, researchers, and engaged citizens from various countries—people who are actively involved in game-changing scientific research. Those people share insiders' insights on the issues everyone needs to know about, such as strategies for replacing King Oil's multi-trillion-dollar monopoly. They intend to see carbon fuels and nuclear fission replaced with clean decentralized alternatives that benefit all of Earth's citizens. The stakes are incredibly high.

Biggest challenge

We have also learned that neither inventors nor investors realize how long it takes to go from proof-of-concept to commercial product. The timeline is especially lengthy for truly breakthrough energy inventions because they lack academic and bureaucratic support. Revolutionary inventions are not born into a simple invent-and-reap-rewards world.

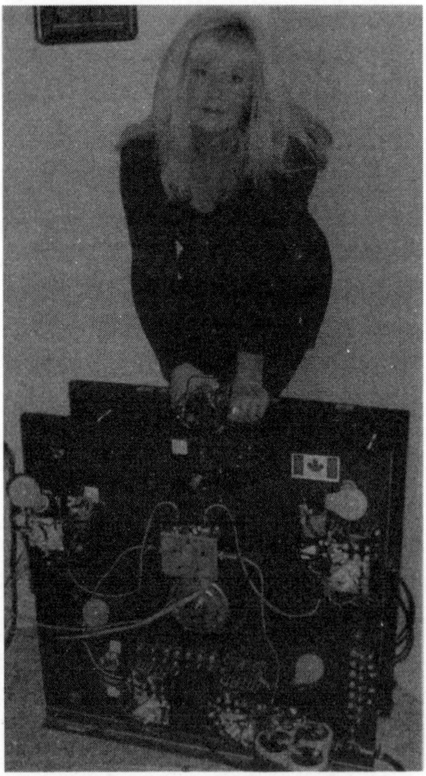

Carmen Miller and her father's magnet motor/generator.

Confronting geopolitics, inertia, fear and greed in order to birth a New Energy Era is the biggest worldly task humans have faced. Lone inventors have believed they personally had to take on the weight of that task and some felt crushed by it. Bill Muller tripped over the need to raise funds for his research. However, after his death in 2004 the next generation carried on the work.

Bill and Ilona Muller's daughter Carmen is determined to see her father's invention being used to help provide abundant clean power for the people. To carry on the task, she teamed with a philanthropic business person to fund the engineering development of her father's technology as well as some other inventions.[5]

Inspired inventors

Bill Muller was one of many inventors we've met worldwide who were nudged by dreams, accidental discoveries or intuition to make breakthroughs in the power-conversion field. One well-informed aerospace journalist has noticed the remarkable proliferation of such inspiration. British author and aerospace insider Nick Cook writes for *Financial Times, The Wall Street Journal* and other major media and is routinely invited to speak to major aerospace corporations, government think-tanks and universities. Years ago he joined the staff of the world's leading military affairs journal, Jane's Defense Weekly and is the magazine's aerospace consultant. While researching a forthcoming book he traveled to Vancouver, Canada to meet with Carmen Miller[6] and learn about her father Bill Muller's inventions.

I interviewed Cook at that time [in September of 2007] and at one point steered toward the question I first encountered in the Muller home: "What sustains the strength of powerful permanent magnets?" These magnets perform work indefinitely—lifting, attracting or repelling other objects, while textbooks say they cannot do work.

Cook agreed that the way magnetism interacts with the background nonmaterial field—zero point energy, vacuum energy or whatever the field is—does seem mysterious. If magnetism triggers a weird or seemingly magical portal for accessing energy from this field, scientists do not yet understand the process. In places ranging from garage workshops to more sophisticated laboratories, people do appear to be accessing the field, but *how* is not completely understood.

Many inventors seem to be building devices intuitively and experimentally but not by following any particular theory, he noticed. From where are these people—in different parts of the world at or about the same time—getting the inspiration and information on how to build the devices?

He speculates that the knowledge is bubbling up into the collective consciousness. The existence of a global-brain Internet and sharing of knowledge is only a partial explanation for the simultaneous inspirations. Something additional may be happening.

"It's a bit like that moment in Close Encounters (the 1977 film Close Encounters of the Third Kind) where Richard Dreyfuss starts sculpting the Devil's Tower out of mashed potatoes."

Cook's analogy is apt. In that film, the actor Dreyfuss plays an Everyman role. Other characters in the film who are driven by an implanted suggestion or vision of the same mountain also come from ordinary or humble life circumstances. Similarly, the inspired, and even obsessed, inventors that the authors of this book have met come from diverse backgrounds. Many had happily led simple lives until the "free energy spirit" whispered its secrets to them.

On the other hand, although relatively few garage inventors have advanced academic degrees, the stereotype of an eccentric who is out of touch with reality doesn't fit the majority of researchers we meet at international conferences. Our friend the astrophysicist Brian O'Leary once commented on the difference between the stereotype and the real people. "I have been surprised to see a breadth and depth of knowledge, dedication and professionalism, and substantial achievements among leading theoreticians, experimenters and inventors in the 'free energy' field. These are the explorers of a new reality. They are cut off from the mainstream because the mainstream debunks this reality—with a denial based on the most superficial reasoning."[7]

The preceding introductory vignettes had to be told in the first person voice because they are solely my experiences, but the remainder of the book is in our combined voice as co-authors.

Our goal

Breakthrough Power presents a sampling of the variety of energy-source choices—for powering our homes, industries and vehicles—available to humankind. The implications of staying on our present disastrous course are presented, as are the boundless opportunities contained in the choice for a bold new direction.

Our conclusion is that it's up to all of us—the citizens of Earth—to make the clean energy abundance revolution happen. Our goal is to help generate the critical mass of human caring and collective will power necessary for creating a wondrous new era for our civilization and our precious Earth.

2

Reclaiming Decision-making

The Stone Age didn't end because they ran out of stones.

—*SHEIKH YAMANI, former OPEC oil minister*

New ideas are resisted. . . . But we must rapidly explore these new technologies, because what is at stake is life.

—*ADAM TROMBLY, Astrophysicist*

There is almost no area of human activity that would not be dramatically affected by the advent of new energy technology—especially matters of war or peace and health and the environment.

www.infinite-energy.com

A time of great changes can bring families together; even a scattered extended family tends to unite when faced with an unprecedented challenge. Some members continue to distrust or squabble, but people of common sense cooperate in seeking a wise course of action.

Despite the intensifying of resource wars and societal breakdown in some sectors of our world, the co-authors of *Breakthrough Power* see higher ideals emerging in the human family as a whole. Ordinary people are recognizing that the unprecedented economic and environmental changes and challenges of our time are not confined to national boundaries. We know people in many countries who are both inspired *and* determined to help humankind and clean up Earth.

Acclaimed author Paul Hawken recently spent ten years examining the worldwide movement for social change and for restoring the environment. He found a diversity of such efforts, ranging in size from a single activist to a billion-dollar-budget nonprofit, that collectively make up the largest movement on Earth.[8] Hawken sees it as unstoppable.

New energy movement

The New Energy Movement (NEM)[9] is part of that grassroots push for a better world– potentially a crucial part. NEM was started in 2003 and based in Portland, Oregon, and in 2007 the allied nonprofit organization New Energy Movement Canada (NEMCA) was formed in Canada, based in Vancouver.

Concerned citizens formed these groups because humankind's unwise past choices of energy technologies have degraded the natural world and the belief that we must fight over scarce supplies of fuel has degraded our civilization. The new-energy activists aim to spread information about some hopeful news. They have learned that some of the new and rediscovered energy-related inventions could turn the situation around—in favor of both the environment and social justice.

There is *variety* of choices for clean, small-scale but powerful, decentralized (localized) very low-cost energy systems. The New Energy Movement intends that the use of breakthrough power inventions will be wise, peaceful, just and in harmony with nature.

We co-authors have faith in the innate goodness, common sense and creative powers of ordinary people who seek the highest interest of the whole of society, rather than a narrow self-interest. Politicians, bureaucratic insiders in the machinery of government and highly-placed men and women in large corporate enterprises naturally have similar innate powers, but appear less likely to champion certain worthy causes if they perceive risk to their privileged positions.

Disillusion in DC

Joel Garbon experienced this situation firsthand. As president of the New Energy Movement, he went to Washington DC early in 2007 after working with colleagues to draft a groundbreaking piece of legislation titled the Energy Innovation Act. NEM executive director Stephen Kaplan had worked for a United States senator and has a long history in public policymaking circles, so he set up meetings with legislative staff of various Congressional representatives during the five-week foray of the two NEM officers. Their goal was to present the draft legislation to members of Congress and have it adopted as the template for a bold, visionary energy policy that aggressively funds and accelerates the research, development and deployment of new and unconventional energy technologies.

'You guys are a gnat'

Joel found that the doors in the halls of Congress are indeed open to citizens. However, he was dismayed at how vested interests in the energy industry wield inordinate influence. He arrived in his nation's capital expecting that legislators would embrace what the voters want. Elsewhere in the country whenever he spoke to a citizens' group, the audiences voiced strong support for public funding of breakthrough energy innovation research. In his words:

> I found I was incredibly naïve in believing our work would be welcomed by our elected representatives. My thinking had been, 'Of course everyone wants this legislation! And you are the servants of the public, right? You want this too, for your own families and your constituents. It's so clearly the right thing to do. And by championing it, Mr. Congressman, you'll leave a proud legacy, you'll be a hero!

However, the two men from Portland, Oregon did not write any checks for anyone's political campaign, so their reception was less than enthusiastic.

One senator's legislative aide was blunt. "You know, Joel, every day

I've got an army of energy lobbyists in here—from coal, oil, natural gas, nuclear power, and now from solar power, wind power, and biofuels. All of these lobbies are coming through here and you guys are a gnat on the radar screen! You don't have any funding, you don't have a constituency, you don't have a presence."

The aide revealed his jaded attitude. "Yes this is nice that you've got a piece of legislation. Yes we'll read through it and see what you've got."

Joel appreciated the aide's candor, but as someone who sees the vital importance of legislation (see Appendix 1) to fund innovative energy projects, he felt frustrated, angry, and indignant at the displays of indifference, arrogance and ignorance he encountered in the Congressional offices. The elder and politically-wise Stephen Kaplan had to calm down his distraught younger colleague.

Joel's eye-opener

Joel recalls, "Steve did his best to counsel me about the political realities in Washington: that my goals were too high, too idealistic, too aggressive. I started to see just how insidious the money culture has become in our government, our corporations. And I felt despair at the profound lack of wisdom in these supposed leaders . . . my mentor Brian O'Leary had expressed heart-wrenching grief over the selling out of America and the senseless suicidal degradation of our planet; he had warned me not to expect anything that resembled courage and vision from our politicians."

Then Garbon resolved, "Power comes from the people; I'm going back to the people. Clearly it's not happening here."

Back to the people

Both authors have been told—by insider sources in a number of countries—that a minority of officials scattered throughout governments, industry, the military and academia know that revolutionary energy inventions exist but are not wanted by those who control world finances because they would disrupt today's corporate and geopolitical power structures. Wall Street benefits from financing costly electrical utilities. Oil and gas drilling and coal mining create personal fortunes. Governments get fuel tax revenues. Even dealing with radioactive waste has become a vested-interest industry. And the oil-war industry has vast sums of money for lobbying politicians.

Up until this writing, no politician was visibly voicing a viewpoint such as: ". . . but the human spirit is capable of figuring out how to reinvent those socio-economic structures. Let's free the energy of human ingenuity!"

The world changed

Not waiting for politicians, an international subculture of researchers with those goals has been interconnecting for decades, even before the growth explosion called the Worldwide Web.

The following anecdote gives voice to some issues which create need for this book. Building a grassroots movement requires acknowledging that different mindsets are firmly set in place and don't change easily.

Sharing worldviews

While bicycling in London, England, an enthusiastic researcher, "Jason", tried to explain the concept of New Energy to a young woman who is a reclaim-the-streets crusader.

People in the activist's neighborhood had staked claim to an abandoned lot and transformed it into an oasis of natural greenery. She was weeding the community garden when Jason started a conversation. It quickly turned into agreement about the need to reclaim stewardship of the "commons"—such as the common air breathed by all life forms on Earth. She was passionate about cleaning up the pollution in the atmosphere.

In his casual laid-back manner, Jason asked if she had heard about "free energy" inventions.

At that point she began to edge away. "My father's an engineer and he always says 'There's no free lunch'."

Jason quickly added, "Well, 'free' isn't the best word because the devices themselves won't be given away. But the *source* of power for radical energy inventions *is* free, just like sunlight and wind. There's a whole variety of revolutionary alternatives most people haven't heard about. 'New Energy' is a better way to say it . . . "

She stopped edging away. "My friends drive a Veggie Car; they fill up the tank with used vegetable oil . . . "

Faced with her pride in her friends' accomplishment, Jason wondered how to tactfully say that burning "biofuels" still uses up oxygen from the air. *Corn, canola or other biomass substances are carbon-based like coal and oil but just not as dirty,* he wanted to say. *And corn should be grown to feed humans, not Hummers.*

"That's better than gasoline . . . " Jason said lamely.

"Or are you talking about hydrogen fuel cells?" she asked, apparently puzzled by his lack of enthusiasm for biofuels.

"No, not the 'Hydrogen Era' as plotted by corporate dudes. Their plan includes thousands of nuke-power plants to make electricity to electrolyze hydrogen out from water. Nuke plants meaning more radioactive junk on this planet. Or else they get hydrogen out of fossil fuels. That's not the answer. The New Energy version on the other hand is innovative 'advanced hydrogen' systems."

"Such as . . . ?"

"Basically, getting hydrogen from water at point-of-use, as in a car. One version is called the Water Fuel Cell. Or there's Dr. Randell Mills' chemical process of releasing the latent energy of the hydrogen atom. . . ."

Although Jason sensed that his intensity for the topic was not appreciated, he rushed ahead.

" . . . And plasma physics holds some answers. There's a physicist in Russia, Dr. Kanarev; he's getting ten times as much hydrogen out of his setup as standard electrolysis and he says the device taps into energy-from-space. . . ."

"I hate technology!"

After her outburst another man would have changed the subject, but not Jason. To connect with her honestly, he figured, he'd have to bridge between their worldviews. "You don't like the result of technologies, I guess. Such as polluting the atmosphere?"

"Right, global warming. Seems like everything we do heats up our poor planet."

"We don't have to. Some types of outside-the-box energy systems actually cool their surroundings while they put out more power than the measured input!"

"Tell that to my dad."

At that point our friend Jason was the one who wanted to end the conversation, remembering frustrating encounters with know-it-alls; her father might be one.

Then he heard the young woman say softly, "Maybe there *are* ways to give us plenty of clean cheap power without depleting anything. But everything powerful gets turned into weapons for killing people." Her expression revealed heartfelt pain.

"That's the way it's been so far," he agreed. "And if we close our eyes and ignore all the good new energy inventions popping up around the world, the weapon-makers keep on taking control. They'll continue to buy inventors' cooperation and continue to grab patent rights and continue to claim that a revolutionary advancement is a 'disruptive technology' —meaning a threat to oil-company profits—and therefore a matter of so-called 'national security'."

Speaking more slowly for emphasis he added, "We're coming up to a big change in thinking in science, but the people don't get told about it. The change will be toward working in harmony with nature, understanding how a hidden background energy spirals through the universe and in everything on Earth and how to tap into that abundant energy."

Her expression showed she was listening, so Jason continued. "At the same time, citizens' groups are forming all over the planet and trying to figure out how to create a more enlightened civilization. What if they added a powerful new energy breakthrough or two, to their collection of tools for change? What if the people got involved in deciding what energy technology their government will support . . .?"

"I get it!" she said. "You want to reclaim science!"

What makes something real?

She had nearly understood his message, but not quite. New Energy advocates respect academic learning and admire those who truly use the scientific method. Although most of the science establishment does resist the energy-from-the-cosmos paradigm that Jason investigates, he doesn't aspire to storm ivy-covered academic institutions and "reclaim science".

Instead Jason was telling her that citizens have a right to influence decisions which profoundly affect everyone's life—such as the choice of predominant energy technologies. Ordinary people, therefore, are wise to learn some easy basics of science in order to watch where energy technologies are heading. Citizens could pay attention to whether the technologies are harmful and chosen for easy profiteering, or if on the other hand they are in harmony with nature and chosen for the good of everyone.

Science of tomorrow

This book is written by admirers of true scientists, and with sympathy for those who've forgotten that to be a scientist is to be open-minded, investigate new findings, do experiments, publish the results for other scientists to work with, and thus learn about the natural world.

New instruments needed?

Some new-energy scientists say that today's dominant science is just not good at detecting and measuring certain types of invisible input energy. They say the finding that today's measuring instruments are inadequate for tomorrow's science does not mean that the input energy does not exist.

As an example, the ultraviolet and infrared portions of the electromagnetic spectrum are just above and below the limits of human vision. Our eyes and brain are quite good at seeing what we call the "visible light" portion of the spectrum, but do not permit us to directly perceive the ultraviolet and infrared portions. Some years ago, breakthroughs in optical instrumentation permitted humans to detect those higher and lower light frequencies, and now products like ultraviolet lamps and night vision binoculars are commonplace.

Is it true that before we could detect and measure these invisible energies they simply did not exist? Of course not. And it is highly likely that there are other pre-existing and universally available forms of energy that science has not yet learned how to detect and measure.

Not every expert resists looking at the evidence for "disruptive" ideas. An increasing number of engineers, physicists and other people with impressive credentials are having a serious look and even getting involved in revolutionary New Energy research.

World-changing energy source?

What is the hidden background energy? We deal with this subject more fully in chapter 11, but let's start with a brief explanation. For most of the twentieth century, science thought of space as being empty. It is not. Space—both interplanetary space and earthly space—is incredibly dense with energy, a sea of energy. This sea of energy fills everything, including our own bodies. Therefore, we can't sense it, nor can we measure it against something else. But there are inventors who say that they have been able to harness this energy, to pull it out of the air and put it to work, without pollution or fear of scarcity.

As wonderful as it sounds, hidden energy is not our only new-energy option. There's a non-polluting seemingly nuclear reaction that can take place on a tabletop. There's hydrogen, a clean fuel that can be extracted from water more efficiently than ever before. There's heat technology, which turns waste heat into electrical power. There's low-impact hydropower, which can tap the energy of our rivers and oceans without dams and flooded valleys. And there are other new energy possibilities.

What are the implications?

A switch from a world economy based on fossil fuels to one based on abundant, clean new energy would dwarf any other event of our times. The leaders of nations and businesses and politicians would be disoriented as they move from the familiar oil-war mentality to an unfamiliar situation in which there would be abundant power for all. In comparison to the crumbling of the fossil-fuel worldview, the fall of the Berlin Wall would be a small blip in history.

We believe humankind is in a new-energy breakthrough period, with inventors developing revolutionary energy devices that could power ships, homes, aircraft, greenhouses, and industries. This power can also be used to desalinate seawater, irrigate deserts, and help fuel a massive environmental cleanup. Why isn't every environmental activist researching quantum leap energy breakthroughs and speaking out in favor of them? The answer involves the roles of the media, academic experts, the inertia of an established mindset and fear of energy abundance. In this chapter we'll briefly look at the history of attitudes toward abundance.

What does abundance mean to you?

When long grasses rippled from horizon to horizon in the prairie breeze of eighteenth-century America, pioneers of that time believed that nearly limitless resources were theirs for the taking. A man only had to plow up the prairie and claim his homestead, then begin to eye neighboring properties. Or for wealth that glittered, the legends said, he only had to roll the wagon farther west and dig nuggets out of the ground in California. If that didn't work out he could backtrack, stake a trapline in the mountains and save profits from the fur trade toward owning a ranch. Buffalo or any "critters" in his way could be taken out with gunpowder. And whenever a pair of lumberjacks sawed through a massive redwood or cedar in the Pacific Northwest, they affirmed their brand of faith in abundance-from-nature.

In the twentieth century, civil engineers tamed mighty rivers, locking the waters behind huge hydro dams and then releasing torrents into electricity-generating turbines. Salmon were cut off from their spawning grounds, but "there are plenty of other fish in the ocean" was a common saying. Driftnet factory ships had not yet begun fishing in that ocean.

Near the middle of the twentieth century a government-subsidized industry claimed it would provide "power too cheap to meter." That (nuclear fission) power turned out to be power too expensive to finance, for Washington state residents. The Washington Public Power Supply System lived up to its acronym "WPPSS!" when only one of five planned reactors was built before the project went bankrupt. The slogan "power too cheap to meter" also left a metallic taste in the mouths of people who had to deal with problems of radioactive wastes.

Myths of the frontier eventually faded. Dams silted, rivers wilted. Asphalt ruled. Today, even in relatively unpopulated Western states, the overflow from metropolitan regions is putting pressure on water supplies and the land. In other parts of the West, the logging of enormous swaths of forests—and the accompanying road-building—ruins mountain streams. Nuclear-fission, coal mining and coal-bed methane energy industries damage rivers, soil and air quality.

Abundance of fresh air everywhere, untouched land and pristine rivers are now a dim memory held by our grandparents, but the frontier mentality of exploit-it-before-the-other-guy-does continues today to steer decisions of those who control large amounts of financial power in the world.

'Abundance' repels enviros

Most people no longer believe in the pioneers' vision of abundant resources, and scarcity realities of the twenty-first century are experienced by conservationists as cause for emotional pain. Many conscientious environmentalists are repelled by the word "abundance," automatically associating it with irresponsible consumerism and plundering of Earth's resources.

In the context of grassroots frustration, insensitive enthusing about the potential for energy abundance usually elicits an annoyed retort. "We have to conserve."

The authors believe the human family also has to *choose*.

The people we speak with at the recycling depot or organic juice bar are for the most part not looking at the *difference* between harmony-with-nature technologies and exploitative practices such as mountaintop coal mining. "Destructive" was yesterday's technology of choice. As a result, the words "science and technology" are repugnant to many of the people who passionately care about health, peace, justice and the biosphere. Usually these acquaintances hadn't heard about the variety of constructive yet powerful clean energy technologies that have the potential to gradually replace oil and nuclear industries if allowed. Wastewater-into-energy technologies could clean up waterways and other variations solve the problem of polluting feedlots and landfills.

In the next chapter you will meet a member of the informal worldwide network of independent inventors of New Energy devices and systems, and others who have the courage to try to change our world for the better.

3

Breakthroughs Needed

A planet is at stake. If you have the resources to help launch a new energy era, do it, and don't be so concerned about what percentage your return-on-investment is.

—*Joel Garbon to venture capitalist [Ashland, Oregon, 2007]*

The people are powerful. But the people are asleep.

—*John Bockris, PhD physicist*[10]

One of the most compelling reasons why humankind needs to change its polluting-power habits can be stated simply: kids on puffers. Puffer is the street word for an inhaler device used to ease the symptoms of asthma. If you see someone having a severe asthmatic attack and unable to breathe in enough oxygen, you won't soon forget the sight of their struggle. Shouldn't the causes of that life-threatening disease be attacked, not just its symptoms?

Asthma rates are climbing throughout the Western world, and preschool children in smog-blanketed cities are hit hardest by the rising rates[11]. Here's the cause of that asthma: Tiny bits of material called "particulates", from burned diesel fuel, grab onto mold or pollen spores which we also breathe in without noticing. By riding on the particulates, those pollutants are sucked deep into our lung sacs and cause the lungs to be more likely to suffer allergic reactions. That means anyone living near a route for diesel traffic is more likely to develop asthma. A Harvard study found that one in four children living in Harlem is asthmatic, and the worst concentrations of asthma are along bus routes.

The problem of particulates from burning of fossil fuels could be wiped out by removing political obstacles to serious funding of research-and-development for new energy systems.

Meanwhile public relations firms are paid to tell people what they want to hear. They parade carbon fuels through full-page advertisements, cloaking the fuels with dishonest praise such as "nature's choice", "green" or "the clean gasoline". Do billboards or magazine advertisements—whole pages of blue sky and green grass—convince you that corporations are going to somehow clean the air with their gasoline?

The advertising phrase "clean coal" is another contradiction in terms. The advertisements ignore the mercury which lodges in our bodies as a result of coal-burning power plants spewing pollutants into the air. Even if scrubbed from state-of-the-art smokestacks, the polluted ashes can't be just taken away and dumped, because there is no "away" location where contaminated ash politely disappears. And even if the carbon dioxide is pumped underground, burning coal—gasified or not—is not an environmentally sane choice. To get at the coal in the first place, atrocities against the Earth are committed; just ask the Appalachians whose mountaintops are detonated to get at a seam of coal. The debris slides into formerly clean rivers and valleys.

The nuclear power industry has been spinning its own set of illusions for years in books, articles and advertisements convincing decision-makers that a producer of radioactive wastes is "clean power" just because its waste product is not carbon dioxide.

If the word-weavers repeat it often enough and long enough, it seems people begin to believe that dirty is clean. The audience for the ads reacts sheepishly, "Who am I to question experts?"

The following section introduces an independent thinker who does question. He is a persistent and courageous example of the new generation of truth-seeking New Energy researchers.

Self-taught scientist advances new energy

While he was growing up around Lake Oswego, Oregon, Graham Gunderson's mental gifts didn't yet have a focus—except to learn all he could about science. Starting at age six, he habitually spent his weekly five-dollar allowance at the electronics store to make electrical circuits—and eventually build projects such as lasers. While in high school, he was a co-instructor in an advanced electronics course at a nearby community college. When he was a teenager trying to impress a girl, he even figured out how to construct a radio station at home.

Graham's family noticed his technical brilliance and insisted he get a university degree. Instead he chose to continue educating himself with the help of science mentors, because he had a strong aversion to anything that could limit his thinking. Why go through an academic system that hammers limiting messages—dogma about what's impossible—into students' heads? he asked.

After high school the spirit of adventure inspired him and a carload of friends to relocate to another city. Picture eight guys in a two-bedroom apartment, with one bedroom full of Graham's electronics experiments. That couldn't last. His friends followed him into his next rented lodgings, where he set up pirate radio station KCUF. Its call letters cheerfully announced the youth's irreverent attitude.

"What about 'Peak Air'?"

While he was still a teenage rebel he found his life's focus while walking to his day job in a yogurt store. Every day he had to walk up a steep heavy-traffic street where vehicle drivers are forced to "step on the gas" for the climb. High stone walls enclosed the sidewalk and roadway into a canyon filled with vehicle exhaust fumes. Since burning fuel takes oxygen from the air, Gunderson found breathing difficult by the time he reached the top.

"You took my air!" he accused the drivers. After work he ranted over the radio waves, mixing humor with his outrage against society's polluting choices. He intuitively knew there are better ways to power vehicles and to generate electricity.

He decided to go on the Internet and learn about the most promising alternative choices, which led him to becoming an experimenter in building potential energy-generating devices. In an online technical discussion group his intelligent comments caught the attention of a researcher who paid him to relocate to California to evaluate and develop a certain alternative-energy invention. Gunderson was only twenty years old at that time. His honesty eventually ended the contract when he told his employer, "You're wasting your money on this (specific approach to energy)." By then the young man had made helpful connections with other researchers. He returned to the Pacific Northwest.

From 2003 on, his focus was on discovering and developing processes that *amplify* electromagnetic energy. A company named Magnetic Power International (MPI) hired him as their head of Engineering Development and he was able to buy laboratory equipment and rent a workspace of about 2,500 square feet.

One day he attended a public hearing on energy-related topics, and waited for a chance to talk about truly innovative energy alternatives. However, people attending the meeting seemed more interested in the timeframe for the end of cheap oil production than in hearing about game-changing discoveries. "Peak Oil" was the buzzword at the meeting. The phrase refers to the time when world oil production peaks. It means cheap-to-pump oil becomes scarcer while global demand for fuel keeps on increasing. Prices for oil and oil derivatives such as gasoline, diesel, asphalt, plastics, and chemicals shoot upward. Peak Oil experts conjured images of ugly battles over oil supplies and the prospect of either a dog-eat-dog declining society or a miraculous widespread return to simple lifestyles of distant past generations.

The focus at the meeting Gunderson attended was not on replacing an oil-based world economy with one based on clean abundant-energy technologies that are harmonious with nature. Nevertheless, when his turn came to make a comment, he told about his technical search for the best breakthrough that would help solve the Peak Oil problems.

Then he asked, "What about Peak Air?"

It turned out that there was little public interest in urging scientists to study the declining percentage of oxygen in the air we breathe around carbon-fueled engines, much less get enough people interested in unconventional energy research to support it. That lack of support did not stop Gunderson, however. The man is dedicated to developing an energy-amplifying discovery he made.

Aladdin's lamp

When co-author Manning met Graham Gunderson in 2007, he was working alone except for the weekly conference call with MPI. The company's other inventor who could have collaborated in the laboratory is across the continent in Florida.

After Manning signed a non-disclosure agreement, Gunderson explained his unique approach to the energy problem. Even as a non-expert, she could grasp the brilliance of his energy invention. Basically he is exploring the interactions between magnets and sound.

Magnetic Power announced that the company is developing a breakthrough technology based upon Gunderson's discoveries. The company calls the system GENIE™ (*Generating Electricity by Nondestructive Interference of Energy*). The goal is to build generators that work continuously, without fuel, extracting electricity by converting the abundant, renewable, extremely dense energy source that exists everywhere in the universe. No pollution is created in the process.

"Variations will provide a permanent power supply that eliminates the need for batteries of all sizes," an MPI press release said. "The cost of electricity from GENIE promises to be less than any competing form of power generation today or in the foreseeable future."

"GENIE can turn future cars into decentralized power plants, can rapidly be commercialized and made in many electronics factories..."

Without mentioning Magnetic Power Inc. by name, MPI associate Lee Felenstein told an interviewer from the electronics industry that the company has "gotten fairly far" in energy-related discoveries toward getting power from static magnetic fields. One of the ways to access electric power is by varying the magnetic properties of a coil's core materials, and the company was finding that a well-known textbook "law"[12] does not apply in all cases.

Felenstein was being interviewed because he had won a prestigious award for his own achievements in the industry[13], but he seemed more excited about MPI's search for a new power source than talking about his own impressive career. The interviewer[14] wondered why more research and development engineers at the government, corporate and college levels were not exploring a power-generation solution with as much dedication as MPI. Felenstein replied that, instead of exploration, those engineers were focusing on product development of existing technologies.

Working lean and alone

Having left his 'impetuous kid' years behind him, Graham Gunderson is getting married the month this is being written. In a way he is also married to the energy project. His piano collection gathers dust, and the song-writing hobby that provided balance to scientific pursuits when he was younger is neglected. Such long work hours would not be as necessary if he could hire even one assistant. Shortage of funding means Gunderson has to build tools and equipment instead of purchasing; he was soldering a circuit board during our most recent phone interview.

We profiled this dedicated innovator at length because his situation is mirrored in that of others who have new-energy gifts to offer humankind. While their friends chat about people met on the job, the lone inventors don't have that camaraderie—much less a budget that allows them to buy the tools they need. Men such as Gunderson don't complain, they just double their efforts. However we believe the genie of inspiration might be heard more often if the inventor and the company supporting his efforts had enough funding.

Race against time

If a clean-energy genie gets into the marketplace soon. Gunderson and countless other citizens who resent inhaling pollutants from burning of hydrocarbon fuels would breathe easily.

In his 2007 book *Energy Autonomy*, distinguished author[15] and German Bundestag (Parliament) member Hermann Scheer makes the point that the people of Earth have been lulled into a false sense of security. The euphoria in the suddenly popular renewable energy community clouds the fact that the influence of the established energy corporations continues to grow.

Scheer, who chairs the World Council for Renewable Energy, says the energy establishment has scarcely changed its deep disregard for renewable energy. Instead, consumption of fossil energy is growing significantly, outstripping the rate of growth of use of renewable resources. And the nuclear power industry is gearing up to populate Earth with nuke plants. Scheer quotes a satirical remark from a Polish wit, "It's true that we're on the wrong track, but we're compensating for that shortfall by accelerating."[16]

Nukes are not the answer

Plans for building fission reactors are certainly accelerating, with the endorsement of a few well-known environmental leaders. The nuclear lobby has for years been funding a public relations campaign, pervasively putting out the false message that "we have no other choice; we can't avoid building nuclear power plants." Nuclear power is trumpeted as the savior, saving the world from greenhouse gases and saving economies from end-of-oil disaster.

The authors have the opposite perspective—that building nuclear plants and creating more radioactive waste is disastrous for our grandchildren, and nuclear plants have a history of being prohibitively expensive.[17]

More than a half-century ago, the late Austrian naturalist Viktor Schauberger (see Chapter 7) warned technologists, "You do everything the wrong way." Schauberger pointed in the opposite direction—away from technologies that create heat, noise and pollution by burning oil, coal and other fuels and exploding atoms. He pointed the way toward energy systems that operate quietly and without friction by using nature's inward-spiraling and suction movements that increase order instead of disorder.

Unwelcome wake-up calls

If we urgently need more blunt words like Schauberger's today then we owe thanks to Hermann Scheer, whom Time Magazine in 2002 recognized as one of five "Heroes for a Green Century". He is one of the few politicians—or authors—who has world-class credentials and honors, media recognition *and* the courage to point out that global conferences have generated more hot air than actual changes. In addition, he doesn't hesitate to push past popular heroes who are afraid to seriously confront the fossil and nuclear Goliath.

Scheer was a speaker at a 2004 climate conference at an estate in Italy. He listened with appreciation to a speech by a famous advocate of climate protection. The speech brilliantly defined climate threats and included charts that indicate coming catastrophes. But what was the cause of these frightening scenarios? The speaker blamed population explosion, affluent lifestyles and technologies that demand more energy use. The speech was followed by thundering applause.

Scheer was dissatisfied with the analysis; it wouldn't change the way humankind operates. Most citizens of Earth don't see what *they* can do about population explosion, expansion of the global vehicle fleet and the life-style choices of other people. As a result, citizens feel helpless. Why didn't the speaker name the key culprit—the use of fossil fuels—and admit that fossil fuels could be replaced by renewable systems?

It was Scheer's turn at the podium next, and he did address that core issue. Climate-conference dignitaries' reaction to his speech was divided, he recalls. Some criticized him for being too concrete, too much of a direct challenge.

New energy seen as competitor?

Joel Garbon has experienced similar reactions when challenging the latest wave of groupthink.

At one conference in late 2007, a large audience of movers and shakers in the West Coast venture capital and "angel investor" communities gathered to hear the good news in the conventional renewable energy sector. The event focused on investment opportunities in clean technology. It was a day of eloquent speeches, glitzy presentations, and articulate commentary on these positive trends.

Speaker after speaker discussed solar power, wind turbines and biofuels, emphasizing the steep growth in investment capital directed toward those technologies. Climate change was acknowledged repeatedly, as were the well-worn observations of the escalating energy appetites of China and India. Yet there was no mention of possible solutions to the looming global crises; the focus was on profit opportunities.

Finally Garbon felt compelled to break his silence and stood to address the crowded room.

"Certainly the new investments in the *conventional* renewable energy technologies are an encouraging sign of increased public awareness of the need for clean energy. But any serious analysis shows that solar, wind, and biofuels are woefully inadequate to get us where we need to go and as fast as we need to get there."

"While capital pours into the conventional technologies, visionary scientists and inventors researching breakthrough energy technologies are languishing due to lack of funding, ignorance and indifference at the political level—and in many instances, official obstruction and even heavy-handed oppression."

"I encourage each of you to become educated about these unconventional emerging technologies, and bring your resources to bear in launching the new energy revolution that our planet so desperately needs. We must explore all options seriously and urgently!"

He was prepared to answer questions and share good news about some of the promising new technologies, but a silence had fallen over the crowd. Many in the audience seemed to shift uncomfortably in their seats. And before any question could be publicly addressed to Garbon, the conference moderator rushed to the microphone and promptly dismissed the session with no commentary on the issue that had just been raised.

As the crowd broke up, a handful of individuals—some skeptical and some enthusiastic—asked him for more information. But most of the attendees kept what seemed to be a wary distance from him. Even the conference organizer, previously a friendly acquaintance, reacted coldly to him after he spoke out.

Clearly Garbon's brief remarks were perceived as disruptive to the majority group-thinking at the conference, just as in Scheer's case. Perhaps some of the clean tech gurus felt their status as experts threatened because his comments exposed their ignorance of breakthrough energy research.

Have people been fooled into thinking that black is green, or been bullied or censored by vested interests? Until the public consciously admits the extent to which spin-meisters have muddled everyone's perception of the energy scene, then anyone who seriously challenges energy monopolies—whether carbon fuels, nuclear, or conventional renewables—will continue to be a lone voice.

What is leverage?

A lever is a tool used for prying or dislodging something, so what does "leverage" mean when we're talking about new energy systems?

Here's an analogy: Think of standard renewables such as solar, wind, biofuels and geothermal energy as a standard-sized crowbar being used in an attempt to pry loose the heavy foundation of the international energy sector and raise it to a higher level—less pollution.

Then realize that breakthrough inventions would be a comparatively huge and effective crowbar that could overcome the inertia of the polluting sector, move it out of the way and allow clean energy systems to build a new foundation-economy for the world.

Outmoded energy choices

Products based on petroleum, natural gas, and coal will be useful to humankind for a long time to come, so these industries will remain far into the future. However, for the good of all life on earth, we must move away from their wasteful and dangerously polluting use as fuels in the transportation, heating, and electricity-generation sectors.

Nor are nuclear fission power plants a responsible choice while we have so many clean options. These power plants have created a legacy of radioactive byproducts, forcing future generations to keep watch over countless tons of our dangerous wastes which include hundreds of tons of plutonium stockpiled around the world.[18]

The problem gases

A network of hundreds of respected scientists, the International Panel on Climate Change (IPCC), has warned that humans are causing climate change by the increasing load of polluting gases we send up to our atmosphere—so-called "greenhouse" gases that trap solar rays and cause a heating effect on the Earth. Like glass on a greenhouse or car windows, the gases allow the sun's rays to enter but not to exit that easily. The result is heating of the air, land, and oceans.

Carbon dioxide (CO_2) and methane are the greenhouse gases of biggest concern. It's common knowledge that burning of carbon-based fuels such as coal, oil, gasoline, diesel, natural gas, and wood gives off carbon dioxide. You may not know about another large source—the cement industry. Cement kilns release CO_2 when they convert limestone[19] into lime.[20] And the intestines of feedlot animals and waste-digesting bacteria in treatment plants and landfills are large sources of methane. Methane is the main component of natural gas, and large amounts escape to the atmosphere during oil drilling and coal mining.

Some climate-change scientists talk about a tipping point—the point of no return regarding the melting of Earth's ice and other dire effects—and say that to avoid passing the tipping point, the human family needs to reduce its greenhouse-gas emissions by 80 per cent by the year 2050. Others speak in terms of urgent action being needed in the next few years.

Biofuels: no magic bullet

When those scientists began speaking out, the public relations campaign for "biofuels" heated up with a flurry of glossy advertisements. Biofuels include ethanol made from corn and sugar cane, and "biodiesel" made from soy, canola, and palm oils, and from waste cooking grease. Biofuels are made from plant materials and generally burn cleaner than the gasoline and petroleum diesel fuels they replace.

Since biofuels contain carbon, they also emit carbon dioxide when burned. The plants used to make biofuels are said to absorb an equivalent of CO_2, so advocates describe biofuels as "carbon neutral".

With politicians pushed by the lobbying and campaign contributions of large agri-business interests who crave profits from biofuels, legislation has given billions of dollars in tax subsidies to spur the booming corn ethanol business. We believe that is a mistake, because the use of biofuels creates its own set of problems:

- Huge acreages of corn are being diverted for making vehicle fuel instead of for feeding people. The result is that people who are already suffering financial poverty have to pay more for food.

- Growing corn and processing it into ethanol is not carbon-neutral. Oil is used to make pesticides and herbicides sprayed on the fields, diesel fuel powers farm equipment and vehicles that transport the crop to the refinery, and natural gas is used to make fertilizer. Natural gas also powers stages of the corn-to-ethanol refining. Some prominent researchers claim that corn-based ethanol consumes more energy-equivalent in oil and natural gas than the ethanol itself yields as a fuel.

Is corn-based ethanol mainly an expensive political gift to agribusiness conglomerates?

Worldwide, we see unintended consequences of the mad rush to embrace misguided biofuels policies. Huge tracts of tropical forests in Malaysia and Indonesia have been cleared to create palm plantations for biodiesel production—destroying unique plant and animal species and eroding fragile tropical topsoil. What appeared to be "green" turned out far differently.

Better than corn or palm oil

Other plant materials would be wiser choices than corn and palm oil. Hemp, castor beans, algae, switchgrass, crop residues, and forest thinnings are alternative feedstocks for biofuels. They offer a large source of raw materials from land unsuited for food crops, and they don't need large inputs of petroleum and natural gas. Lower cost methods of converting cellulose to ethanol are emerging.

In the next section of this chapter we hear from a top scientist who advocates growing algae in seawater for biofuels, while voicing an urgent wake-up call about our world situation.

'Perfect storm'

Dennis Bushnell, PhD, chief scientist at NASA's Langley Research Center, says new, decentralized, advanced technologies are quite capable of saving us from energy-related devastation—*if society seriously supports the new systems fast enough.*[21]

In the scenario Bushnell describes, we're heading into a Perfect Storm of energy-related effects. He points to the following combination:

The era of cheap oil ends, but new customers sign on.

- Billions of people in China and India line up to become consumers.

- Around the world, fuel prices go up.

Inertia glues humankind onto the carbon-fuel death ride.

- Fuel-consuming industries that grew huge in the Industrial Age resist drastic change; the ruling industries don't want fuel-less energy inventions to replace them.

- Labor movements fear revolutionary change because of negative messages hammered at them such as "The economy will collapse" or "you won't have a job".

- Political systems are slow to cut their ties to oil and nuclear energy, especially since the fuel-producing industries supply the big dollars to politicians' campaigns.

- Governmental planners are stuck in thinking that they must depend on fuel taxes for revenues.

- The public isn't pushing for a changeover to a radically different energy technology.

  Earth continues to heat up.

- More and more electricity-generating plants are built to burn coal or gas or radioactive fuel and millions more fuel-burning vehicles are unleashed.

  Methane makes it worse.

- Bushnell's scientific message, put in plain words, is that warming of the Earth's atmosphere will accelerate when ancient methane deposits trapped in Arctic permafrost and frozen at the ocean bottoms evaporate into the sky and add to a greenhouse effect—of gases trapping and reflecting heat back to earth. At the same time, increasing numbers of people in developing countries are starting to use today's technologies which create waste heat.

Seawater algae for biofuel

At a Foundation for the Future conference in Seattle[22] where Bushnell spoke about alternative energy paths, another scientist[23] said our oceans are in trouble. Carbon dioxide is acidic, and seawater absorbs some of the excess atmospheric CO_2. This process is acidifying our oceans—with potentially devastating effects on the food chain.

Dr. Bushnell and other distinguished scientists say the most urgent issue is consciousness-raising of the public. To avoid disaster, we, the people, must learn about both the problems facing us and the solutions. For example, he favors growing salt-tolerant plants on deserts irrigated by seawater in order to grow biomass for fuels without taking prime agricultural land, but he is also open to possibilities such as zero point energy.

The entrenched industries and entrenched politics of Industrial-Age energy systems will only respond quickly enough to "a major public movement and outcry," the NASA scientist concludes.[24]

Freedom in the land of giants

To gain energy freedom, perhaps we need to find the Achilles Heel—the weak spot—of the profiteering giants of polluting energy. That weakness may be the giants' need to control us. If an energy conglomerate aims to keep the people dependent on utility companies and centralized power plants so that the conglomerate gets the maximum amount of money out of us, then a small portable system for tapping into an abundance of universally-found free energy would quickly find that Achilles Heel.

A breakthrough solar energy collector combined with advanced batteries would threaten the giant's control, as would manufacture of other less well-known systems outlined in this book. Tapping into a previously unrecognized source of energy, from the very air around us, is the ultimate in energy autonomy—freedom from fossil fuels. No one will work toward manifesting that positive vision of energy self-sufficiency unless they hold the vision clearly in mind—and in heart.

The next chapter is intended to assist you in helping to manifest that freedom, by envisioning a better future and encouraging others to do the same. If you for some reason find it too "blue-sky" for you, be assured that the chapter ends with the practical although visionary planning of an inventor whose technical work requires no new theories to explain the results. It is grounded in principles that are found in textbooks but have just been ignored.

4

The World We Can Choose

Imagine a time when we will have no pollution, no poverty, where every home and vehicle has its own access for generating power. This is not the Jetsons and futuristic 2300. . . . We have the ability now to create a civilization free from all sources of pollution and free from roads between cities.

—*Dr. Steven Greer*[25]

Vision is radical common sense in action. An imagined goal organizes our intelligence and lights our fire. It brings forth genius and talents.

—*Marilyn Ferguson, author, Aquarius Now*[26]

Contrary to popular belief, there is no shortage of innovative energy technologies. When the New Energy Movement hosted its first public conference, in 2004, guest speakers said the problem is *not* finding clean technologies that could be effective enough to replace polluting oil, coal and nuclear fission.

"Engineering is the easy part. . . . The sky is raining with technology," said tidal power expert Martin Burger. Instead the problem is political. At a deeper level, prevalent attitudes are the problem. Other speakers noted that the theme of the conference, The Courage to Change, was apt.

Can we citizens make a positive difference simply by changing our own attitudes, our vision of the future? There is evidence that the power of coherent positive thinking can indeed help create whatever happens. Imagine having a renewable source of clean power that is available day and night and independent of whether the sun is shining or the wind is blowing. Consider that clean, truly harmless, low-cost and abundant energy is our birthright.

"If you don't have a dream, you'll never have a dream come true." These lyrics from an old musical[27] contain more than a grain of wisdom. Current scholarly studies about human consciousness point to physical-world effects arising from our thoughts, and new physics hints that how we view our world affects what our experiences will be.[28] Experiments such as The Intention Experiment[29] also explore human thought and intention as a tangible energy—an inexhaustible but simple resource that can be used to improve our lives, communities and planet.

If you want to use the power of your imagination to help create a better future, we invite you to ponder the possible impact of wondrous small scale devices that provide abundant energy while working in harmony with nature. What could breakthrough clean energy inventions mean for you and your family?

A healthier world

To progress beyond the common tendency to focus on disasters, consider this: If we tend to create whatever we envision with emotion, then why invest in apocalyptic visions? We believe it's time to picture in glowing detail what a brighter future could look like—envision a higher civilization that powers its life-affirming activities by responsibly tapping abundant energy from the universe.

Picture the region you live in and imagine it thriving under sparkling clean skies. Even a large city in such a world is quieter and the air refreshing. Internal combustion engines have been replaced with quiet electric motors; now you can hear the gentle sighing and singing of breezes and birds. The rumble and roar of traffic have been silenced; even in the heart of the city, sidewalk conversations can be conducted with a quiet voice, and the nighttime atmosphere is calm.

How does fresh air taste when vehicles, buildings and industry have truly non-polluting power plants? Imagine today's asthmatic children breathing freely and growing up with healthy lungs. Picture trees and flowers adding delicate fragrances to clean oxygen-rich air. No longer does toxic vehicle exhaust pollute the air and contaminate roadsides; apples, berries and other nutritious and delicious food can be grown on curbsides and freeway medians.

Picture what a higher civilization would look like, feel like, taste like, if we, the people, insist on wise governance. Without the call for oil wars or environmentally damaging energy megaprojects, public funds are focused on improving health care and housing, building better schools and infrastructure, stimulating creative and life-enhancing work, and restoring our environment.

A world of sharing

Consider being generous with your envisioning. Imagine that, in a time of abundant energy resources, countries and cultures other than your own also prosper. The level of antagonism in the world lowers. Picture the majority of the human family cooperating to cope with social problems that energy abundance alone does not solve.

Cheap, clean decentralized electricity stimulates sustainable local economies and provides the catalyst for a new wave of job creation—meaningful, rewarding life-affirming work that enhances human dignity and contributes to the planetary clean-up and advancement of global society. Low-cost energy permits recycling on an unprecedented scale, and the unsightly gigantic heaps of trash that previously blighted the landscape are transformed into valuable mines of recovered resources.

The world's thirsty families are supplied abundant and pure drinking water extracted from the atmosphere itself or by desalination of ocean water. Depleted farmland is restored to productive agriculture.[30] Streams, rivers, and aquifers are no longer exhausted by damaging diversion and over-pumping, but are restored to their natural balance, permitting fish stocks and streamside ecosystems to thrive once again. Pollution from animal feedlots no longer goes into streams; manure, garbage and other wastes are turned into sources of electricity.

Food can be grown anywhere —even in desert or arctic conditions— in compact, multi-level greenhouses whose natural heating and lighting are augmented by energy easily and cleanly extracted from space itself. No longer do children suffer hunger, and the indignity of starvation on a planet of plenty is banished forever. The idea of wastefully transporting food thousands of miles becomes unthinkable, as every community grows nearby a variety of nutritious food. Gone is the tasteless, waxy food that was genetically-modified to withstand the rigors of long transport and storage; it's replaced with the succulent natural heirloom varieties that can easily be grown and supplied locally.

The vision is not idle dreaming; it is quite possible. Prototypes of inventions that use novel combinations of resonance, magnetism, states of matter, certain geometries or inward swirling motion to unlock the secrets of universal energy have already been built. They provide proof of new or rediscovered principles. In many variations of these inventions, a small input triggers a disproportionately large output of useable power.

These energy converters don't violate any laws of physics if they simply tap into a previously unrecognized source of power—background space. A flow of energy from that source can continue day and night, whether or not the sun shines or the wind blows.

There need be no energy apartheid, no haves and have-nots when it comes to powering homes, businesses and other human institutions by tapping into that abundance. We hope that eventually geopolitics will settle down to acceptance of the fact that everyone in the world is sitting in a sea of energy and it is possible to build devices to extract ample power from it. Many such inventions are in the research phase. While these cosmic energy technologies are being developed, we already have a variety of other clean energy innovations, from geothermal heat pumps to ocean wave power converters to improved electric vehicles.

An environmentally positive vision of the future pictures the damaging resource extraction winding down soon, with its stakeholders shifting their support to the quantum-leap clean energy sector. By sharing the vision, the technologies, the work, and the rewards of a breakthrough energy era, our civilization can collectively step through the door to a world of remarkable promise.

Rethinking the economy

When we talk about quantum leap energy technology, opponents of the energy revolution—and some proponents—warn that "it'll crash the economy." Meanwhile national economies are already on shaky ground and highly vulnerable to the volatility in the fossil fuel markets. In fact, having our world economy based on increasingly expensive carbon fuels is causing economic distress on all levels, from national down to our personal households. The most obvious "price" of costly oil is its immediate negative impact on wallets and budgets when we fill up the fuel tanks, but the less obvious and perhaps much steeper price is the interconnected chain of environmental and geopolitical consequences of policies based on unsustainable polluting sources of energy.

With the specter of severe climate change impacts and resource wars on the horizon, a healthy nudge in the direction of "disruptive" clean energy technology could be the global economy's salvation rather than the harbinger of its collapse.

It is no accident that an increasing number of politicians have as a central tenet of their election campaigns the promise of bringing "green collar" jobs to their home territory. A revolution in "cleantech" centered around new energy technology, would indeed result in the creation of a whole new wave of well-paying jobs.

Imagine the amount of work provided by the need to retrofit all existing internal combustion engines, changing vehicles into clean power plants. Imagine all the work required in installing "home power modules", and decommissioning and removing the millions of miles of unsightly power lines that crisscross the landscape.

Consider the explosion of job opportunities in environmental remediation, as new technologies clean up contaminated air, soil, rivers, and oceans. The family farm makes a comeback, in a world where local communities produce all their own food.

New energy technologies will spawn an endless number of derivative companies with improvements for many sectors—sustainable agriculture, heating and cooling, medical treatments, transportation, industrial processes, and diverse hardware and systems. The economic boom that is catalyzed by new energy will likely dwarf those spawned by the computer, Internet, and telecommunications revolutions.

Restoring forests and rivers

Where the authors live in the Pacific Northwest (Jeane in Vancouver, BC, and Joel in Portland, Oregon), it is all too common to see whole mountainsides clear-cut for their timber, and miles of logging roads snaking through pristine forest. Forest soil exposed by this energy-intensive and expensive method of tree harvesting washes into streams. The eroded soil silts over salmon spawning beds and endangers other aquatic species.

But imagine how both the environment and the forest industry would benefit if new energy technology resulted in a fuel-less helicopter. Suddenly selective harvesting of trees from the air would be economical. No logging roads would be needed. Instead of clear-cutting swaths of forest, heli-loggers could selectively thin the weakest trees. Forests could be restored to pristine "old growth" status.

Without the need for hydroelectric power, there is no need to dam rivers. With certain dams dismantled, salmon in the Northwest would be able to easily return upriver to spawning grounds. The vitality of cool, clear, unimpeded flowing waters could be restored.

Compassionate changeover

While a new energy era can enhance prosperity, health, peace and beauty on our planet, it will not occur without the pain of dislocation. Many jobs in the current polluting energy infrastructure will disappear, similar to what occurred when buggy-makers and office typists became obsolete. Breakthrough advances do cause losses, but generally many more opportunities than losses are created.

An energy revolution would affect millions of workers worldwide who earn a living from the coal, oil, gas, and nuclear industries. Wise and compassionate policies will be required to help them adjust to new modes of work in the age of clean energy. A new attitude—of sharing wealth liberated by energy abundance—can be the basis for a successful transition strategy.

What would happen if the best minds on the planet—in specialties such economics and geopolitics and planning, along with brilliant multidisciplinary generalists—put their heads together and figured out how an alternative world economy would look? One scenario they could consider is that of a global network of self-sufficient bioregions, meaning that people provide food and stewardship of water and other resources within their own region. Locally-generated electric power would be a safety feature—decentralized freedom from the vulnerable central power grid.

The new era won't come overnight, nor will technology fix everything. Aerospace journalist Nick Cook reminds us, "The technology coming out of vacuum energy, zero point energy, spacetime energy—call it what you will—is going to take a while. In the meantime we've got to start changing the way we live."

Bringing a new era into view

Why haven't you heard about the new energy possibilities? Part of the answer is history repeating itself. At the start of a scientific revolution, most experts in the old worldview can't see or acknowledge the new paradigm. William A. Tiller, PhD, a frontier scientist who both experiments with and writes about the physics of the future, tells a story to illustrate how established mindsets can jail us.

After the 19th century naturalist Charles Darwin sailed his ship, the Beagle, from the Atlantic into a harbor of the land he called Patagonia, the sailing ship anchored at a distance and the crew rowed smaller boats to shore. Darwin found that although the sky was clear and native Patagonians on shore could see that horizon clearly, for some reason they could not perceive the sailing vessel. Only their shaman was able to notice the ship.[31] For the next few days the shaman described the sailing ship to his people, relating it to their experiences and whatever memory comparisons they had. Suddenly the Beagle emerged into their view and they all saw it.

Envisioning a new clean energy era, and talking about it to your neighbors, brings the emerging era out of invisibility and into sharper focus. They will then see that there's not one, but dozens of new energy "ships" on the horizon, quite different from one another. The choices include

- non-polluting engines that run on waste water;
- ultra-efficient solar technologies;
- new heat-to-electricity inventions;
- plasma-based energy converters (forget blood plasma; we're talking about the electrically-conductive type of plasma which is the state of matter beyond gaseous);

- resonant electrical circuits;
- low-impact tidal power;
- wind power combined with super-efficient magnetic generators;
- new ways to release hydrogen from water at point of use;
- and many other innovative systems.

Envisioning the physics of abundance

World-traveling physicist Mark Comings told the New Energy Movement conference audience that new physics reveals in a scientifically unassailable way that certain beliefs that profoundly impacted human psychology for millennia are a lie. Scarcity, limitation and lack are those ingrained beliefs.

We do have an array of solutions emerging from the brilliance of the human mind and heart. New energy discoveries are backed by new knowledge that involves a revolution taking place at the foundations of physics, Comings said. You don't see it on the television news yet, but the debate emerging out of new science is about our understandings of fundamentals—space, time, light, mass and force. The public tends to think that physicists agree on the fundamentals. Not so, especially during a scientific revolution.

Fullness of space

Current physics talks about the vacuum of space, but we now know space is energetically anything but empty. He eloquently described what he calls the plenum of space. Plenum means fullness. He views space as being full of infinite energetic potential, virtually boundless luminosity and sentience. Sentience means awareness. Comings found that all the great spiritual traditions characterize consciousness as equal to space.

Comings sees a spiritual awakening on Earth at the same time as old systems falter and kick up a fight before fading away. In Europe, Asia and on other continents, he meets grassroots groups who are evolving a new planetary culture in which loyalty is given to the biosphere, fidelity given to the interconnected field of life.

The awakening is happening despite cultural conditioning that focuses each person on separateness. The separation mentality results, for instance, in some inventors being sunk in them-versus-us thinking. "I can empathize," said Comings, "because there are tyrannical powerful forces manifested in the form of the powers-that-be . . . actively fighting against and preventing the innovations that we are dreaming up and wanting to give to humanity as a gift."

He has in the past performed successful energy-converting experiments, creating a ringing resonance by injecting certain frequencies into piezo-electric crystals. When the crystal was in resonance with the plenum of space, the power output rose significantly higher than the input. He concluded that, if allowed politically, such discoveries could guide humankind in building a completely clean energy infrastructure—resonant technologies that allow us to live in harmony with the universal energy field and the Earth.

Freedom's another word

New energy visionaries don't fit any stereotype; they range from university students to white-goateed nuclear and electrical engineer Dan A. Davidson. He has researched new energy devices all his adult life and has written new-paradigm books such as *Shape Power.*[32] Davidson writes about the probable impact of new energy devices:

> Various power groups know that if mankind has unlimited energy at its disposal, it becomes virtually impossible to control and manipulate people. With free energy a person is not subject to those who would control his transportation via fuel curtailment. One could live virtually anywhere, since a readily available energy supply could be used to make any environment livable. Water could be taken from the air via condensation if necessary; and with water, food could be grown.
>
> With unlimited energy available to a country, it could synthesize anything, including the atomic elements; therefore that country is not open to international blackmail because of energy resource requirements. Stated in brief: energy = freedom.[33]

Solving a dilemma

Scott McKie of Seattle is a down-to-earth engineer and inventor who has for twenty-five years also been working on an energy-generating device that has no moving parts. He distances himself from talk of new science or new physics, saying simply, "My system is not connected to any of the other power sources that you are talking about here, except resonance.

"Resonance is a real 'here today' phenomena that just hasn't been looked at very carefully, and that is because of the (nonsense) that physicists have been putting out about 'energy' for decades. Electronic resonance has been used in radio tuners since their inception, while in the field of electric power resonance is avoided like the plague."

"My system is not 'of the future', but here on this earth with all of the world's present problems." McKie said his real-world solid-state electronic system could be understood today, if physicists were not teaching that it cannot be achieved.

". . . When physicists decide to 'come clean' and say that they don't know it all, we'll all be a lot closer to reaching the gleaming world that you are describing. In the meantime, my project is back here in the dirty old 'here and present' and doesn't need anything of the gleaming futuristic thinking to bring it to fruition."

Having stated that disclaimer, we can tell you that McKie has wrestled with the questions of how to bring his invention into society, and has come up with some answers.

He has a detailed plan for manufacturing his "power-on-demand modules" in a sparsely inhabited region of the Pacific Northwest where a model city could be built to house workers.[34] McKie believes that a planned city needs to be built "because I want to put people to work, not robots." Assembly of small parts could provide jobs for stay at home parents around the world.

He and his business partner who worked for a large utility company figured out how to integrate the modules into existing power plants and local neighborhoods. The win-win business plan would lease the fuel-less modules to utility companies for replacing standard generators. The idea is that utilities could pay off their infrastructure debts and workers won't lose their jobs.

Existing vehicles would be retrofitted with the electronic product. Mechanics would still have jobs—checking each vehicle every few months. Transportation costs for everyone would be slashed by using the invention to power vehicles, even the semi-trailer trucks that haul loads long distances.

Then an apparent roadblock came up while McKie and his partner were figuring out how to integrate their new energy system into society. They realized that government officials would be unwilling to give up fuel tax revenues.

McKie's partner pondered the dilemma. One morning he had an answer. "Replace the fuel tax with a distance/weight tax for truckers and their loads." The state would have revenue and the truckers would still be ahead financially because they would have fuel-less systems powering their trucks—no gasoline or diesel expense.

Who can be trusted?

Meanwhile, decision-makers in national capitals tiptoe around such topics as the implications of unlimited-energy freedom for everyone. Perhaps policy planners discuss the implications behind close doors.

There have been glimpses of insiders' awareness of revolutionary energy systems. In 1986, we were shown a document in which the United States Air Force was requesting proposals to explore the use of energy from the field of background space. The military wanted some defense contractor to research ways to use "esoteric energies heretofore unknown, including the zero point fluctuation dynamic of space."[35]

We expect that if and when government officials do openly acknowledge non-conventional energy choices, their resulting policies will reflect not only vested interests if any, but also their faith or lack of faith in human nature. Can military officials resist the temptation to distort new energy technologies into even more devastating instruments of destruction? Can ordinary people be trusted with empowering technologies that allow living comfortably in inhospitable locations (with cost-free heating and cooling) and travel across continents without fuel costs?

We think such an energy revolution is inevitable. The sooner society's leaders openly plan for it, the smoother will be the transition.

In the next part of *Breakthrough Power*, we meet the heroes of the past who pointed us toward the natural world to learn how to achieve energy abundance—by life-enhancing instead of life-destroying processes.

PART•II

Hidden History Impacts You

When they built machines that seemed to defy "laws" of physics, individuals had their moments of fame, but their work never appeared in school textbooks. Did vested interests block their paths toward mass-marketing radically cheap, clean energy inventions? Or was the human family not yet ready for a paradigm of energy abundance?

The hidden history of new-energy technologies is rich with human triumphs and setbacks. It's a story that impacts real people. If mass media reported it, however, bright individuals worldwide would ask, "What can I do to help?"

Co-author Joel Garbon and other scientists look to past revolutionaries for clues on how to tap into the background energy. The following chapters introduce some of the many little-known heroes who uncovered keys to energy freedom. They each believed in working with nature instead of against it.

5

Nikola Tesla

Ere many generations pass, our machinery will be driven by a power obtainable at any point in the universe . . . . Throughout space there is energy.

—*Nikola Tesla*

Nikola Tesla (1856–1943) is called the Father of Free Energy. Despite his major contributions toward creating the modern electrical generation and distribution systems, textbook authors still largely ignore his accomplishments. The entertainment industry sometimes brings Tesla's name forward in films and novels, but usually misses the crux of his story. We believe it is important to understand that this extraordinary human:

- possessed advanced abilities to access inspiration and information;
- proved his genius by pioneering the electrical system we use today;
- was so devoted to discovery that he was willing to overturn that electrical generation and distribution system for a better way;
- tried to take humankind beyond our crude nineteenth-century energy technology—to life-enhancing knowledge about tapping into a universal energy field;
- learned some characteristics of that energy field through hands-on experiments and found ways to coax what he called its "radiant energy" into machinery.

Complex celebrity

The elegant inventor Nikola Tesla was a favorite of New York's high society in the late nineteenth century, and he often entertained celebrities at his laboratory. Friends such as Samuel Clemens, better known as author Mark Twain, stared open-mouthed at the sparks roaring from Tesla's special electrical transformer. At times, their host would stand in the spray of high-frequency electricity while a glass tube in his hand lit up with no wires attached. And in hotel dining rooms and private parlors, Tesla's creativity and intellect attracted other cultural stars, including author Rudyard Kipling, pianist Ignace Paderewski and naturalist John Muir.

Tesla was cool and detached, yet charming, a man of contradictions. Although a loner, he was also a showman. Slim and tall, always perfectly dressed, he commanded attention. His most striking feature was magnetism, a combination of dark good looks, intense blue eyes, and an aura of mystery. It seemed the world was destined to be his.

It didn't happen that way, despite his having given the world a complete system for using a rotating magnetic field—the electrical equivalent of inventing the wheel. By the time Tesla died at age eighty-six, his inventions and theories had been largely forgotten by the public or seemingly discredited, and his plan to provide free energy worldwide discarded.

Financial hardship and powerful opposition helped cause his downfall, and later other innovators who admired Tesla would run into similar problems. Despite ridicule from "experts" who have not paralleled the rigors of Tesla's work, in the 21st century relentless researchers and experimenters are still uncovering Tesla's secrets.

Meet the real Tesla

Television scenes portraying Tesla as a mad scientist or man of lightning promote an inaccurate stereotype. Only a few books tell the true complex story. For instance, from childhood onward Nikola Tesla deeply appreciated the natural world and arrived at many of his breakthrough discoveries through studying nature's ways.

As a young man he noticed unhappily that scientists were headed toward a view of the scientific method that mainly deals with material measurements. The mechanistic view is more geared toward measuring quantities than studying qualities, and tries to reduce living nature to mere forces and mechanisms. A biographer, Gerry Vassilatos says Tesla knew that worldview would eventually lead to violent responses from nature itself.[36]

Amazing abilities

Nikola Tesla had extraordinary perceptive abilities. He began life in the eastern European region then called Croatia, son of a Serbian minister and a mother who was an inventor in her own right. As a child he claimed to be awakened by the crackling of distant flames, and alerted his parents to save their neighbors from house fires.

If Tesla had not been such an extreme sensitive, he may not have developed the system of electrical power generators and motors we use today. He was inspired to solve a major science puzzle by suddenly and literally visualizing the solution.

It happened long before he moved to the United States. While still in Europe the young scholar/engineer from Serbia wrestled with a challenge—how to come up with a new principle so that engineers could build efficient electrical machinery. The direct-current (DC) system used then was inefficient but society seemed stuck with it. Experts of the day did not see how alternating current (AC) could replace DC. Tesla intuitively knew there was a better way, and he clashed with a professor who scorned his ideas as "impossible".

From breakdown to breakthrough

While intensely concentrating on the electrical-current puzzle and other challenges, Tesla was hit with what his doctors called a nervous breakdown. His senses became painfully acute, more so than ever. While suffering from visual as well as auditory oversensitivity he developed an unusual ability. Any invention he was thinking about he could see as clearly as if the device were solidly in his hands, or as a detailed hologram that he could walk around.

After recovering from his "breakdown", Tesla took health-restoring walks with an athletic friend. According to Tesla's account, one afternoon they were walking in a city park during a luminous sunset. It reminded Tesla of lines from his favorite poet, Johann Wolfgang von Goethe, who was also a natural scientist. Tesla spontaneously recited the lines:

> The glow retreats, done is the day of toil;
> It yonder hastes, new fields of life exploring;
> Ah that no wing can lift me from the soil,
> Upon its track to follow, follow soaring.

Suddenly he was transfixed by the setting sun and a vision of a vortex whirling in space. The elusive answer to the technical problem came like a flash of lightning. Tesla could see a stunningly simple motor operating in front of him. It was an entirely new approach—a rotating magnetic field producing alternating electrical currents that are out of phase with each other. Tesla saw a workable AC system that would revolutionize electrical technology. He tried to convey his excitement, "See my motor . . . Watch me reverse it!"

His friend of course couldn't see the vision and must have thought Tesla had really lost his mind.

Tesla grabbed a stick and drew a diagram in the dust. Six years later, in 1888, he would show a similar diagram to the American Institute of Electrical Engineers and tell the world about his new scientific principle involving the magnetic whirlwind created by out-of-step currents. The professor who would suggest the lecture spoke out and said Tesla had more than a new motor; he may have the foundation for a new technology and its essence was Tesla's elegantly simple induction motor which had almost no moving parts to break down.[37]

Edison versus Tesla

While still in Europe, for a time Tesla worked at Continental Edison in Paris. His boss Charles Batchelor wrote a letter about his employee to the famous inventor Thomas Edison, "I know two great men and you are one of them; the other is this young man."

When Tesla arrived in the United States in 1884 with little more than twenty dollars and that letter of introduction in his pocket, Edison was already a wealthy, powerful man. Tesla won Edison's grudging respect soon after Edison hired Tesla as an assistant. Working eighteen-hour days, seven days a week, Tesla conquered difficult technical problems. Tesla on the other hand admired what Edison had accomplished by trial and error and with only a grade school education.

Edison soon lost his industrious new assistant, however. Tesla had described how he could improve the efficiency of Edison's generator, and Edison had clearly replied, "There's fifty thousand dollars in it for you if you can do it." After months of work, Tesla did significantly improve Edison's machine. But when Tesla asked for his money, he was shocked to hear Edison say, "Tesla, you don't understand our American humor."

Edison wouldn't pay. Tesla walked out.

Rather than work for a man he didn't respect, Tesla dug ditches on a New York street crew. Three years later, his luck changed. Financial supporters gave him the chance to have a laboratory and develop his polyphase system of alternating current. He designed and patented a new motor, generator, transformer and other necessary parts. Industrialist and inventor George Westinghouse of Pittsburgh bought all of Tesla's patents on the system and signed a contract to pay Tesla start-up cash and stock plus royalties of $2.50 per horsepower produced by the system.

War of the currents

Now the clash with Thomas Edison heated up. Edison fought the development of AC electricity because Edison's lamps ran on direct current (DC), which is a flow of electricity in one direction. DC can be sent for only a few miles on power lines. In contrast, alternating current flows back and forth in a regular rhythm. It easily travels for hundreds of miles down high-voltage lines.

Edison didn't want to hear about those advantages. He had a lot of money invested in his system and saw AC as a threat to his business. His strategy in the War of the Currents included electrocuting dogs and publishing scare pamphlets, all in a public-relations attempt to link AC with death. But despite Edison's efforts, Tesla and Westinghouse won.

Westinghouse built an AC system for lighting the 1893 World's Fair in Chicago, and Tesla was a star at the exhibition. In white tie and formal jacket with coat-tails, and wearing cork-bottomed shoes for protection from electrical currents, he shared a stage with one of his Tesla coils—a device he invented that generated very high-power currents. Its bolts of electricity crackled and snapped and lit lightbulbs in Tesla's hands.

The crowds loved the drama, and the exhibition's success led to development of a hydroelectric project at Niagara Falls. Eventually Tesla's distribution system delivered immense amounts of electrical power across the continent. Since the contract with Westinghouse gave Tesla $2.50 per horsepower, he should have been wealthy for life.

They ran into a problem. George Westinghouse was in financial trouble, with business competitors trying to squeeze him out of the power picture. Tesla remembered that Westinghouse had believed in him when others hadn't, so Tesla tore up the contract, took a cash settlement and walked away from the millions of future dollars assigned to him by the per-horsepower deal. Although Tesla enjoyed money when he had it, it was more important to him to see his friend's company survive.

Tesla's loyalty to a friend was extraordinary at a time when fortunes were being amassed by giddy entrepreneurs hoping to join the "captains of industry" who rode the wave of the second Industrial Revolution. Tycoons prepared to make fortunes in utility companies. They wanted the AC system to cover the earth with power poles and wires. Power companies would eventually dam rivers and pollute the skies with noxious exhaust from coal-burning power plants.

In contrast, Tesla wanted to create an even newer system—of energy that would be transmitted worldwide for free. His proposal was "free energy" in the sense that it would have been sent at no cost to the consumer except for the cost of a tuned receiver. Experimenting toward that goal led him toward an unusual "open system", in contrast to the closed-path electrical system which had become accepted and today is used worldwide.[38] Unlike the War of the Currents, this was a war that Tesla could not win. This time the big money was against him. Power tycoons did not want free energy for anyone who simply stuck a receiver into the ground, and opposed any other energy system that can't be metered and billed.

Tesla planned to send out messages—similar to what we now know as radio—and energy, both without wires. The energy part of the plan was radical enough to eventually cause Wall Street to slam doors in his face. It was a time when power monopolies would soon be floating in money; no one wanted to rock the boat. Corporate moguls such as banker J. Pierpont Morgan had already bought up copper mines. It did not take much insider information to figure out that transmission wires would someday cover much of the world with nets of copper strands.

As if oblivious to the schemes of monopolists, Tesla went on to propose his startling new idea—the worldwide transmission of free power.

Tesla plan: broadcast power to the people

In the same year, 1893, that he dazzled society by lighting the World's Fair, he gave a speech at a prestigious institute in Philadelphia and talked about earth resonance—part of his vision for wireless power. It involved sending out electrical pulses of the correct frequency through the earth to create special waves of energy. Just as a piano string will vibrate when another instrument at a distance hits the same note as the pitch to which the string is tuned, a Tesla receiver or antenna would be tuned to resonate with whatever the Tesla transmitter broadcasts. Such resonance would fulfill his dream, expressed in an 1897 speech, of sending power from station to station without using any connecting wire. He saw a day when such a system would both speed communications and supply limitless energy.

An ordinary person would have been too distracted by fame to think about such things, but Tesla was not ordinary. In the next few years he wrote and was given patents on processes for futuristic wireless transmission of power and messages, despite the chance that they might make his own previous inventions obsolete. His mind leaped ahead and his experiments progressed far beyond standard electricity.

Tesla was not the first inventor of alternating current, but he did invent the complete polyphase AC system of motors and generators we use today. Inventor William Stanley independently built an alternating-current power plant and lit up the main street of Great Barrington, Massachusetts, a year before Tesla designed his Niagara Falls station.[39]

What then was Tesla's greatest discovery? Some researchers say his most stunning claim is not even considered by the mechanistic science worldview. The claim involves a different type of electricity, with different qualities. Standard AC electricity is potentially harmful to living beings, in comparison to the quite different "cold electricity" that is being researched by frontier scientists.[40] *Nikola Tesla was the pioneer of such experiments.*

Laboratory in cow pasture

To test his new ideas, in 1899 Tesla relocated to the mountains near Colorado Springs, Colorado. In a high-altitude pasture, he built a high-voltage laboratory. Picture a building wrapped around the world's largest Tesla coil, with a strange pole placed on top. There in the shadow of Pikes Peak he worked toward his new goal of sending special vibrations throughout the earth and tapping the universal energy field.

Tesla-researchers are still deciphering the accomplishments of his months in the mountains. He wrote sometimes-terse notes, keeping in his head much information about the invention's operating principles. His existing notes have to be translated into today's electrical engineering terms. But Tesla legends feed on the facts of his Colorado Springs experiments with unprecedented power levels and frequencies. Several millions volts pulsing at 50,000 cycles per second is a challenge even today.

When he tuned up his massive coil, fifty-two feet in diameter, he could create a twelve-million-volt discharge and throw bolts of more than a hundred feet in length from the copper ball on top of his flagpole. The townsfolk kept their distance after rumors flew around that the famous inventor could make lightning. Meanwhile, thunder from the electrical discharges reverberated for at least fifteen miles.

While working in uncharted electrical territory, increasing the voltage and current dangerously high, one day he stood in the doorway of his laboratory and watched the copper ball at the top of his pole while his courageous assistant was at the switch inside. Explosions of lightning roared above the building, and the inside of the cavernous building filled with a strange blue light. Suddenly the apparatus created a huge unexplained surge of excess power that backtracked on the power lines, over-rode safety circuits at the Colorado Springs Electric Company power station and burned out its generator. Local authorities were so angry that when their standby generator went online they refused to send electricity to Tesla's laboratory. This incident proved costly to the mysterious scientist both in terms of the expensive repairs needed and in damage to his reputation. Yet he was undeterred, and convinced that his experiments had proved the viability of amplifying the power of electromagnetic waves.

Generations later the world had occasion to remember that unexplained surge of power. In the 1970s, Russian scientist Aleksandr V. Chernetskii, PhD also burned out a power station with a surge of power that he attributed to cosmic energy.

How could Tesla send electric power without wires through the ground? William J. Beaty recently figured out how to explain that within conventional physics. (Extraordinary Technology article gives his full explanation.) Beaty says the conductivity of Earth's ionosphere and a certain (dielectric) effect of our atmosphere make Tesla's wireless power system feasible. Tesla had noticed that the Earth resonated when struck electromagnetically—"ringing" like a bell—but at first he didn't realize why Earth's atmosphere is also crucial to his system.[41] *That's not a big mistake when compared to that of scientists who insisted the Earth doesn't have resonant frequencies. Since they didn't understand how it could work, they dismissed Tesla's wireless scheme. Beaty adds that when Very Low Frequency earth resonance was rediscovered in the 1950s, nobody in conventional science circles openly admitted that Tesla had been correct.*

On June 7, 2007 MIT announced results of an experiment by scientists[42] *who sent wireless power a short distance (up to five meters, or across a large room) to power a light bulb. That limited distance was wrongly cited by skeptics as proving that Tesla's wireless transmission could not have traveled the distances he claimed. The MIT experiment involved resonance of components; but not resonating the earth. Tesla's experiments did involve the earth.*

Wardenclyffe and the fall of Tesla

Tesla returned to New York in January 1900, satisfied that he knew enough to carry out his vision of wireless transmission of electricity. He hired an architect to design a 154-foot-high wooden tower on top of a brick building on Long Island. It was to be used as a huge transmitter, with a mushroom-shaped copper electrode on top. Tesla named the project Wardenclyffe, envisioning it as a station to broadcast wireless power as well as worldwide radio communication channels.

The tower was nearly finished in 1902, along with the 100-feet-square building intended for a powerhouse and laboratory. Then perhaps at the height of his creative genius, Tesla's dreams began to unravel.

Tesla's vision of sending communications through "wireless intelligence" had convinced financiers such as J.P. Morgan to pay for the research, but they didn't realize that he also intended to send free wireless power to people everywhere. Tesla had left out that part when he spoke to Morgan, his main backer, in 1900 about financing for Wardenclyffe. Tesla had instead pitched the opportunity for Morgan to take monopolistic control over all radio broadcasting, but the tycoon would give Tesla only limited funds. Three years later, desperate for money, Tesla admitted to Morgan his true intentions.

Morgan had investments in power-related industries, and in any case was not known to give things away. He cut Tesla off. Free power to the people? No way.

The radio broadcasting opportunity was lost as well. Marconi beat Tesla to that prize by sending a signal across the English channel. Tesla frantically tried to both find another investor and develop commercial products to pay the bills. In 1906, construction on Wardenclyffe stopped. Eleven years later the tower was broken up for scrap metal, long after Tesla lost his mortgage on the place. The expensive adventure taught Tesla that powerful electromagnetic waves could circle the earth and resonantly build up strength. However, it was only one of the liberating paths that energy technology could have taken as a result of his experiments.

By the time Nikola Tesla died in 1943, his great achievements of the 1890s had been largely forgotten, and some science writers ensured that he was mostly remembered for private eccentricities such as fear of germs. University students are given the impression that he invented the Tesla coil and had a unit of magnetic measurement named after him and that's about it.

A backlash from the academic community may have contributed to Tesla's descent. He did not play the game; he had no interest in getting articles into academic publications. Tesla's showmanship—his public displays, like the one at the 1893 World's Fair—may have caused professional jealousy. Historians Oliver Nichelson and Christopher Bird said of Tesla: "So advanced were his concepts that science and industry of his day were unable to comprehend their essence and scope."

Resurrection

Modern day interest in Nikola Tesla has heated up. The Internet is now rich with shared technical information on Tesla's theories and patents, and nearly every major bookstore harbors a Tesla biography. The questioning is intense. Did he really run a car on "free energy"? What did he mean by "radiant energy"? Or his magnifying transmitter?

Recently the evidence that Tesla could pump cosmic energy from our surroundings has been multiplying, with revelations about what Tesla called radiant energy and how it relates to inventions such as a "cold electricity" motor[43] or the work of inventor John Bedini (see chapter 17).

What has society lost by ignoring or dismissing Nikola Tesla's later ideas? Could it be that humanity could now be enjoying the benefits of a clean, low-cost, universally available and *safer form of electrical power?*

In the next chapter we meet another pioneer, one of many who have looked to Tesla's work as the forerunner of their own inventions. This next pioneer lived in the American West—and even drew gunfire.

6

Surging Sea of Energy

The universe is singing and this symphony of frequency is what keeps every part of the universe and every atom in its proper orbit.

—*T. Henry Moray*

The electricity coming from (T.H. Moray's) device had a strange "cold" characteristic.

—*Moray King*[44]

T. Henry Moray (1892–1974) of Salt Lake City, Utah, was a boy tinkering with crystal radio sets when he discovered the writings of Nikola Tesla in 1900. Tesla had said that a form of energy pervaded the universe and if this energy was in motion then it could be used to generate power. The boy took Tesla's words as a challenge and started on the path to becoming an electrical engineer who pursued a dream—that humans could mine energy from the cosmos by stimulating and amplifying existing oscillations in space.

In pursuit of cosmic energy

By age fifteen, Henry had a job wiring houses and the beginning of his own Radiant Energy[45] idea was flashing in his mind. In the summer of 1909 he started experimenting with taking electricity from the ground. By autumn of the next year, he had enough power to run a miniature arc light. He firmly believed in his idea, despite accepted theories which would not have embraced it.

Two years later he knew the energy he was working with was not static, but instead it oscillated (swung back and forth). He also believed it was not arising out of the earth, but instead was pulsing here from some outside source. His experiments indicated the electrical oscillations or "ether waves" surged like waves of the sea, pounding the earth more by day but also through the night. What was the source of these perpetual vibrations? Henry concluded they came from a colossal reservoir of energy out in space.

After taking a correspondence course in electrical engineering, he traveled to Sweden on a mission for the Church of Jesus Christ of Latter Day Saints. The young missionary took time to study science at the University of Upsalla and complete a doctoral thesis on the idea that there is energy throughout space.

Mysterious white stone

In the summer of 1913 at Abisco, Sweden, Henry took some soft, white stone-like material from the side of a hill, tested it and decided it might be useful in a valve-like detector of energy.[46] The stone evidently contained very weakly radioactive material and this maintained plasma activity in a valve.[47] From this stone the young missionary began to develop the "Moray valve" that was used in his early Radiant Energy devices.

When he returned home he worked at jobs such as design engineer for a large electrical switch yard. An industrial accident in 1920 injured Henry's eyes, and as a result of being unable to work at his profession he returned to Radiant Energy research. To make mortgage payments, he bought 3,000 chicks and began to raise poultry. He was a family man, a respectable citizen and listed in the 1925 *Who's Who in Engineering*. Senator Reed Smoot heard about Henry Moray's inventions and that summer invited him to a meeting in the legislator's office in Hotel Utah. Moray offered to give his Radiant Energy invention to the United States government. According to the account in *Sea of Energy*, the senator graciously declined; the government would not accept such an offer—because "the government was not competing with public utilities."

By then Moray had spent countless hours in his basement working on his special valve that detected radio frequencies, a key part of his invention which pumped free energy out of the supposedly-empty "vacuum" of space. Moray's explanation for the source of the electricity: "Energy can be obtained by oscillatory means in harmony with the vibrations of the universe . . . the Moray Radiant Energy Device is a high-speed electron-oscillating device."

Free energy demonstrated

A typical demonstration given by T.H. Moray in his earlier years is described in the book *The Sea of Energy In Which The Earth Floats*. One of the witnesses later wrote that when the demonstrations and congratulations were over, he felt confident that Moray had a genuine power-generating invention and that no hoax was being perpetrated. Moray's explanation for the source of the electricity: "Energy can be obtained by oscillatory means in harmony with the vibrations of the universe . . . the Moray Radiant Energy Device is a high-speed electron-oscillating device."

In other words, Moray was pumping "free" energy.

In the 1930s, his Radiant Energy Device worked for days at a time, converting what he saw as cosmic energy into 50 kilowatts of useable electricity, enough power for 25 average homes. The tabletop device with no moving parts produced a strange cold form of electricity which ran a motor, heated a flat iron and lit bulbs. A considerable current could be guided on thin wires without heating them.[48]

Witnesses and trouble makers

Prominent people in the 1920s and 1930s witnessed many tests of Moray's device, and the press was satisfied that the tests were valid. However it seemed that the Radiant Energy Device was not wanted by some people, as Moray received numerous death threats. Even so, Moray repeatedly showed his strange electric generator to creditable witnesses. The only threat that stopped him from further demonstrations came in the form of legal advice from his patent attorneys in Washington D.C. Under patent laws he could have lost his rights to a patent if he continued to show his invention to just anyone.

The U.S. Patent Office was not much help; it rejected seven patent applications for Henry's device because the generator and its output were outside the bounds of the conventional understanding of physics at that time.

Moray's first germanium-based device was submitted to the U.S. Patent Office in 1927, and rejected on the basis that it would not work without a heated cathode. Henry Moray was so far ahead of his time in semiconductor technology that the patent office said his invention was impossible. (Society later learned that cold cathodes are possible.) Despite his pioneering use of semiconductors, when the transistor was officially invented 20 years later no credit was given to Henry Moray.) Nor did he get what he wanted most—the satisfaction of seeing his invention precision engineered for mass production to help humankind.

Gunfire in the laboratory

T. Henry Moray pointed the way toward a revolutionary change in the way we convert energy, and for his visionary efforts he drew fire—not only verbal shots, but an actual gunshot attack resulting in a bullet to his leg. A man named Felix Frazer who had been sent by the Rural Electrical Administration to work in Moray's lab went crazy with a sledgehammer (some reports say an axe). The man destroyed the Radiant Energy machine.

Henry's son John Moray said "the hazards of fame" came to the Morays in the 1920s. "My mother received phone calls telling her that her husband's life was 'not worth a plugged nickel' unless he cooperated on Radiant Energy . . . His home and laboratory were constantly broken into when the family was not at home."

John remembered an incident in 1936 in Salt Lake City when the children were in the family car, with his mother driving, and he was in the back seat. A bullet went through the car and lodged in the windshield in front of his mom. "A classic black sedan with all the shades down almost forced her off the street, then sped away up 21st South."

Someone unknown shot at Henry's friend who was the first president of Moray's Cosray Research Institute and the bullet fortunately whizzed past and out of the friend's car. In one attempt to break into the Moray house, someone shot the Morays' watchdog. Henry bought two handguns to protect his family and he had bullet-proof glass installed in the windows of his automobiles.

John Moray was 13 years old in 1940, and as a grown man[49] he chronicled his memories of the harassment that the family experienced. As a result of the constant threat to his life, his father carried a small gun in his pocket, and whenever he walked from the house to the laboratory at night he wore a revolver.

"On three different occasions, he was attacked at his laboratory and shot his way out of the situation . . . "

Moray said he believed the harassments were intended to force him to turn over his laboratory notes to Felix Fraser and associates from Russia. A congressional investigation later indicated that Communists had infiltrated the Rural Electrical Administration. From then on, trouble multiplied. Henry Moray refused to cooperate further with the REA. Many of his records disappeared. His family later discovered that more than a dozen of Henry's original patent applications had disappeared from the U.S. Patent Office, although the file jackets remained there. Who stole the dozen or more patent applications? John Moray said that question will probably remain unanswered.

T.H. Moray continued with his dream of harnessing a usable source of free energy that could benefit the world. Toward the end of his life Moray became increasingly dismayed, as it had been his firm belief that his work could not be stopped, and yet it had been de-railed by persistent and heavy-handed opposition.

The amount spent by the Moray family on the Radiant Energy Device was the equivalent of millions of today's dollars, considering how difficult it was in the 1920s for a family to scrape together $200,000 for materials and equipment.

When Manning interviewed sons John and Richard Moray, John told her that Radiant Energy could have been made available for commercial use because his father's final devices were closer to that level of refinement; they needed no antennae or ground wire. Henry's sons were still hoping to locate an investor who would fund the expensive final research-and-development stage for the Moray device—standardization of the components.

Some researchers believe that T. Henry Moray's secrets died with him and that his sons will never be able to replicate his device because Frazer's sledgehammer had destroyed the proprietary and expensively crafted parts of the Moray device.

Human nature has not changed, Richard said pessimistically; forces of greed will probably prevent such an invention from being mass-produced today just as it was stopped in the past. "But do not misunderstand," he added. "I want the world to have Radiant Energy. Radiant Energy is factual. My father did have Radiant Energy," he said with apparent strong feeling.

Betrayal by respected expert

Conversation turned to scientists who were prestigious in Henry Moray's time, and who witnessed the Radiant Energy Device but later denied having seen it work. One eminent scientist particularly harmed T. Henry Moray's reputation by denying publicly that he had seen a working model of Moray's invention. Richard Moray and a relative visited the scientist when he was well into his nineties and apparently making peace with his life.

"He admitted that, yes, the Radiant Energy Device worked, just like my father said." As Richard Moray revealed this, lines indicating deep frustration etched into his face. He continued slowly. "I asked him 'then why; why did you do what you did?'"

The scientist's excuse for denying Moray's invention credibility? In a flat tone Richard quoted the dying man. "He said, 'because I couldn't admit that I didn't know'."

The scientist had a reputation as an expert but he did not know how or why Moray's invention worked. Rather than admit that, the man had lied and betrayed T. Henry Moray.

New generation of engineers catch the vision

While T.H. Moray was nearing the end of his life, a youth named Moray King was attending school in Utah. When King finally encountered the book *The Sea of Energy* and noticed the author's surname was identical to his own unusual first name, Moray King paid attention because it was such a strange coincidence that he also had been researching zero point energy. He had decided his purpose is to explain that energy so clearly that scientists and engineers would be able to create machines similar to those of T.H. Moray. King noted that T. Henry Moray had discovered the importance of ion oscillations, which King sees as one key to tapping zero point energy.

King says, "I was amazed that I could find support for his methods in today's physics literature that seems to explain the device. The fundamental operating principle arises from a surprisingly simple hypothesis: abrupt, synchronous ion surges in plasma appear to coherently activate the zero point energy."[50] T. Henry Moray's device could *not* have been tapping into standard electromagnetic fields, King realized, because it performed in places that were shielded from such fields: a mineshaft, an airplane and a submarine.

In T.H. Moray's lifetime, the evidence was compelling that a revolution in energy extraction was about to begin. However, similar to a Greek tragedy theme, human weakness, insecurity, and greed trumped the life-affirming potential of T.H. Moray's contribution.

The energy pioneer introduced in the next chapter asserted that we approach technology in the wrong way. The correct way would be to work in harmony with nature's ways.

7

Living Energy

At the time I was still unaware that in water the greatest secret lay hidden.

—*VIKTOR SCHAUBERGER, Austrian naturalist*[52]

How could we have missed this universal machine? Why have we ignored the vortex, the workhorse of the universe?

—*WILLIAM BAUMGARTNER, vortex scientist*

The true history of human civilization will begin with the end of energy shortage. To achieve this, however, we need a completely new understanding of what energy really is.

—*DR. HARTMUT MULLER*[53]

In the mid-twentieth century, the Austrian naturalist Viktor Schauberger (1885–1958) studied how hidden energy from the cosmos enters into nature's spiraling motions, including those in flowing water, in moving air currents, and other natural spirals. The result of his observations included energy-harnessing inventions that pointed the way to harmless power extracted from inward-spiraling vortex motion, but most people missed the cue.

Viktor Schauberger and the spiral of energy

As a forester after World War I, Schauberger had spent countless hours watching vortexian turbulence—three-dimensional spiralling—in the water of wild rivers. His employer at the time was an Austrian prince, and the royal family owned a huge area of untouched wilderness. At first there were no roads, dams, logging sites or other human intrusions to interfere with the balance of nature in the vast hunting reserve. As the prince's forester, Schauberger had the benefit of years of study of life processes as he hiked through the forested mountains under his care.

According to his biographers, Schauberger saw unusual sights in this pristine ecosystem, such as a landlocked lake renewing itself with a whirlpool followed by a giant waterspout. At nights, by a waterfall in the light of a full moon, he learned about the heightened energy state of cold water by seeing certain egg-shaped rocks float.

The theme he saw in nature's movements and designs was the vortex. Keeping in mind his motto of "understand nature, then copy nature," the observant genius made what he called "living machines." Today's main energy technologies use outward-moving *explosion*, such as fuel-burning and atom-splitting. By contrast, Schauberger's machines operated on principles of inward-spiraling movements of *implosion*. In short, he had discovered how to generate electric energy in a radically different way by working in harmony with nature's creative movements.

One of the turning points in his understanding of energy took place one day when he startled a large trout in a swiftly flowing stream. He had been wondering how the fish could remain motionless in fast-moving water, with only slight movement of its tailfins to maintain its position. And now he asked how and why did this trout flee upstream instead of letting the current help push it downstream?

The observant forester eventually figured out that the fish's shape and motions caused vortices to form and push the trout against the current. He also discovered a relationship between water's temperature and its ability to form vortices. As an experiment he had his assistant pour hot water into a creek upstream from a trout, which caused it to lose its position and be swept downstream. The vortices that supported the trout were destroyed by the warm water (possibly due to the reduction in water viscosity as temperature is increased). Schauberger's further measurements showed that water temperature drops where the stream shoots around egg-shaped stones. Later he studied how the egg shape interacts with and induces the formation of vortices in many ways. The egg shape is nature's way of creating strength and enhancing vortexian movements within and around that shape.

Schauberger asked other questions that textbooks didn't consider, such as "How can a trout flee upstream in defiance of a strong opposing current? How can it leap and swim up a sixty-meter high waterfall?" Schauberger described the trout as rising through the waterfall's mist and up against the weight of water.

Copy nature

Over time, Schauberger's attunement to nature's ways led him to develop concepts about levitation forces in water. This viewpoint resulted in the unusual machines he built. Some had twisting pipes that had variations of an egg-shaped cross-section. These shapes increased an inward-spiraling flow inside those pipes that compressed and accelerated the fluid in increasingly tight braiding toward the center.

His knowledge of water led to him building a prize-winning flume, or wooden channel, for the prince to transport logs that were heavier than water. Knowing that rivers meander and loop from side to side for good reasons instead of flowing in straight lines, he made his flume snake down the mountainside. Hydraulics[54] experts said his unusual winding flume would never work, but heavy logs rode high in the water all the way to the bottom of the mountain. Again Schauberger's knowledge of the energetic structure of flowing water was proven in practice.

Word spread about Schauberger's approach to hydraulics, and the Austrian government assigned a professor to find out what Schauberger knew. Professor Philipp Forchheimer was astounded to find that although Schauberger's theories contradicted established theory, in practical applications they worked brilliantly. Forchheimer was a true scientist, willing to examine paradigm-disturbing evidence. He even admitted to his students that he had been teaching "rubbish" about hydraulics for the previous forty-five years.

Schauberger's breakthroughs in understanding levitation forces began on a cool moonlit night one spring, early in the years when he hiked alone in the unspoiled forests. As he sat beside a waterfall he noticed a large fish darting back and forth in the river in twisting motions as if building up energy. Suddenly it disappeared up into a huge jet of falling water. He caught a fleeting glimpse of it spinning wildly under a cone of water and then floating upward until it tumbled over a curve at the top of the waterfall.

No scientist could explain this at the time. Schauberger alone realized that even while gravity's pull on water creates a visible downstream flow, invisible levitation currents are going in the opposite direction in a river in its natural state. The forester's developing understanding of the roles of temperature and motion combined with his study of the vortex; he knew that water has its most potent energetic structure in cool and dark conditions.

To imagine an "antigravity force"—in Schauberger's terms, how a flow of energy could pull anything upstream against the weight of gravity—picture the tunnel in the middle of a vortex swirling down a drain. With increasing suction it drags things downward into the gurgling drain. Schauberger, in teaching how a levitational counter-current works, suggested imagining such a whirlpool *turned upside down*. A trout would appear to be floating upward in along the axis of vortex spin. Schauberger said that with the right lighting it is possible to see the path of what he called levitation currents—as a tube within the misty veil of a waterfall.[55]

Schauberger's unusual concepts such as levitation currents flowing upstream in unspoiled rivers are not stated in this chapter as proven science fact. Co-author Joel Garbon, accustomed to using technically accurate language, was inclined to change this section to be consistent with his own knowledge of fluid mechanics and hydrology. (And surface chemistry; the fact that temperature affects viscosity and vortex formation was likely unknown by Schauberger but well-known today). Garbon and Manning decided to present Schauberger's original concepts accompanied by this acknowledgment that aspects of his observations could be explained in scientific terms. For example a fluid itself could be structured to move an object upstream and counter to the pull of gravity.

Implosion generators

Schauberger quit his job as royal forester when his employers began to log the forest greedily instead of selectively, and he regretted having built the innovative flume. We presented the naturalist's experiences during those years in the forested mountains because his firsthand observations of processes in unspoiled nature led to an understanding of principles he used later in his "biotechnical machines." Those inventions ranged from a copper plow for agriculture to an implosion generator for powering a house.

Scientists who had spent an equal number of years learning from books rather than direct observation generally didn't see the value of Schauberger's self-education. Nor did they expect workable energy generators to be based on implosion and vortices, because as academic scientists they were immersed in conventional energy concepts based on explosion and expansion, as occurs with the combustion cylinders and pistons of internal combustions engines.

As he continued to experiment, academics prepared to expose the "uneducated" Schauberger as a fool who didn't know the laws of physics. Instead it became obvious that he had made advancements in science and engineering. He built unorthodox water pipes designed to create forward-moving vortices with the fluid. Because the water spiraled inwardly toward the center and pulled away from the pipe walls instead of pushing against them, water could move faster and with less friction through his pipes.

What this means for us is that a new "free energy" science is possible. No energy had to be added to the system inside Schauberger's unique twisted pipes, yet water was sucked forward with increasing speed in a seemingly frictionless flow. He based his subsequent energy inventions on suction instead of pressure, implosion instead of explosion, compression instead of expansion.

One of the most dramatic moments in Schauberger's research occurred when an associate disobeyed him and made an unauthorized test run on a machine in his absence. We won't describe the circular model except to say that as usual Schauberger had imitated nature's spiraling forms. In this case the gaps between the turbine and base-plate were whorls, mimicking the corkscrew shaped antler of a certain deer species. After it spun up to 20,000 revolutions per minute, the machine called a Repulsator ran itself without the starter motor. Schauberger believed a reaction between excited air molecules and the body of the machine would create a levitating effect.

Schauberger's 1941 letters to contractors indicated he had two purposes for building the prototype: to validate his theories of levitational flight and to investigate production of "free energy."[56] Apparently his associate was too eager to make the test run, and started it up while it was tied down in a hangar. The machine developed so much lift that it broke the cables anchoring it to the floor, shot up against the roof of the hangar and was destroyed in the collision.

Viktor Schauberger's life story is dramatic material. Aerospace journalist Nick Cook wrote a gripping account of the politics surrounding Schauberger's levitation experiments.[57] Schauberger's other biographers cover an attempt on Schauberger's life and the later era when Hitler forced him to head a team of imprisoned engineers. That antigravity project extended until the end of World War II.

Why Was Schauberger's Work Stopped?

In 1958, when he was seventy-three years old, two Americans persuaded Viktor and his son Walter to go to the United States. The Nazis had forced Viktor to work on his energy-generating device in a prison camp—or else say goodbye forever to his family. Now a consortium was promising to manufacture his beneficial energy devices. It was something that he had always wanted.

That visit to America turned out to be an ordeal in a sweltering Texas summer. An atomic energy expert came down from New York, met for three days with the Schaubergers, and reportedly wrote in a document viewed by them that Viktor was likely correct in his projection that his biotechnical innovations were the path of the future. But the Schaubergers' hosts soon revealed their insincerity; they were not in any hurry to develop his generator.

In order to be returned home, Viktor had signed a contract during his stay in the United States that forbade him to ever write about or even talk about his past or future discoveries. The consortium now owned all the rights to his implosion-generator secrets. When father and son stepped on an airplane to go back to Austria that fall, Viktor was broken in spirit and Walter was filled with bitterness toward the United States that lasted throughout his life.

On the way home, Viktor cried repeatedly, "They took everything from me, everything. I don't even own myself." Five days after they returned home, he died, heartbroken. Instead of being rewarded for his work, Viktor Schauberger's life ended in despair.

Why would anyone trick Viktor Schauberger into thinking that they would bring his knowledge out to the world, only to bury the knowledge once they had it?

'Copy Nature'

Schauberger warned that society's fire-technology (exploding the atom for power, and motors that operate by burning and explosions of fuel) is a destructive path. Explosion-based technologies create heat, friction, noise and wastes. Burning Earth's oil reserves, destroying wild rivers with hydroelectric megaprojects and splitting atoms in nuclear reactors also bequeaths havoc to the next generations.

Nature does have a breakdown cycle that involves heat and decay—fire and composting—but nature uses the opposite principles for enhancing life and rebuilding. Schauberger showed how to switch technologically from using explosive to nature's implosive—inward-spiraling—motions. His suction-turbine for instance used a rediscovered ancient principle. "Understand nature's ways," he said. Implosion-based or vortex technologies work silently in comparison to today's technologies. Instead of heating, they more often cool materials, and, if water is part of the energy-converting system, some inventions even vitalize the water.

Schauberger's specific inventions have proven to be more difficult to reproduce than other variations of new energy systems. However, his philosophy of working with nature instead of against nature underlies the efforts of many of the inventors you will meet on the science frontier.

Life enhancing technologies

The air or water inside his implosion-based energy converters was said to be revitalized instead of devitalized. His discoveries also mean we could be drinking healthier water from municipal water pipes, but that aspect of applying his knowledge is beyond the scope of this book, as are his paradigm-changing insights on forestry, agriculture, and in general how we should be dealing with water on this planet. (Vortex principles are now commonly applied in technologies for cleaning water, ranging from removal of grit from pulp slurry at paper mills to removal of solids from wastewater and stormwater.)

His understanding of the processes of temperature change and the two types of electromagnetism found in nature, ordinary and diamagnetism, were crucial to many of the technological improvements he discovered.[58]

With ecosystems degenerating on Earth today, we believe the human family needs to learn from Viktor Schauberger how to restore water to its natural life-force functions. In the opinion of his followers, dam-building must stop because it obstructs the formation of complex vortexian patterns found in the swirling flow of natural rivers, impedes fish migration, damages the quality of water and leads to widespread degeneration in surrounding ecosystems.

Fortunately, there is a healthy opposition to new dam-building, at least in the United States. Nearly 300 dams have been removed from streams and rivers in the U.S. since 1999, and more dam breaches are planned for key salmon-spawning rivers in the Pacific Northwest.

Electrical generation

Wherever new hydroelectric dams will be built and old ones remain intact, ecosystem damage can be minimized by using Schauberger's breakthrough turbine. He built and patented[59] a small turbine that used vitalizing centripetal (inward-spiraling) motion instead of conventional centrifugal motion. It needed only ten per cent of the volume of water that a conventional turbine would need to generate an equal amount of electricity.

Carrying the torch

Schauberger's gentler, centripetal-flow, turbine design could be widely employed today if hydrologists and other decision-makers were aware of its value, but his more advanced energy converters seem to require a research and development task force. New energy researchers are cooperating internationally in building prototypes of his "free energy" designs, but find it a difficult challenge to get implosion motors to work as well as had the original models.

In Canada, William Baumgartner has worked with "applied vortex mechanics" for thirty years. Each time he constructs a new prototype he gains new insights on practical uses for vortex mechanics. He has taught workshops about the creative processes of nature and the universal background mechanics of nature—the transparent universe.

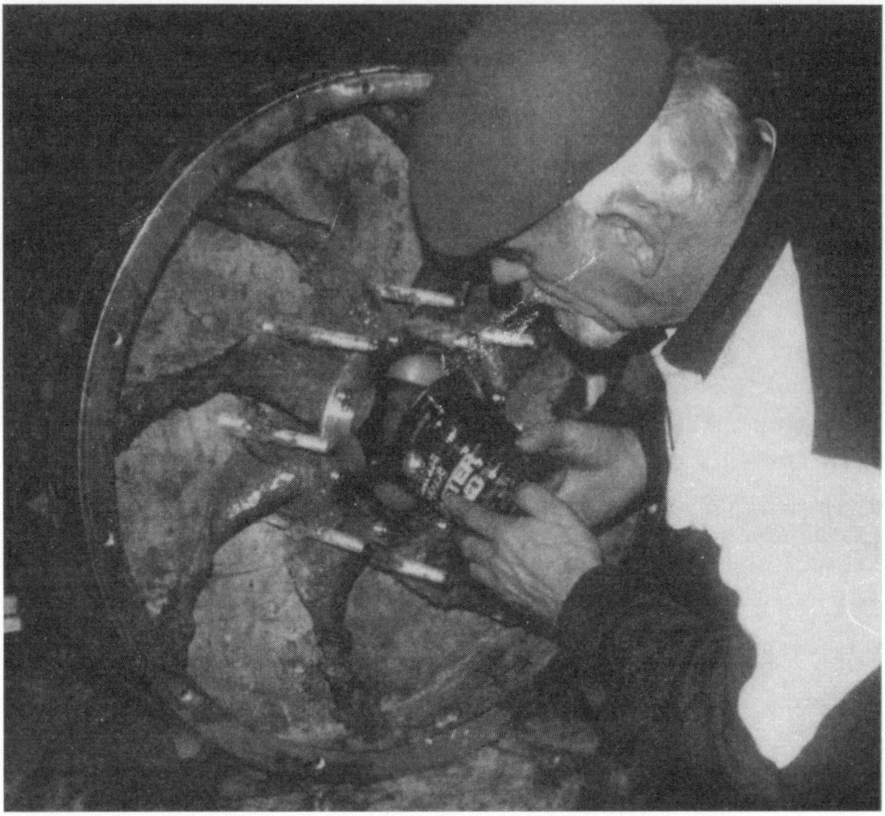

William Baumgartner making Schauberger-type generator parts.

> There actually is an invisible universe, and it's in charge of the visible universe. We can learn how Nature manifests its creations by understanding the background space geometry. Once you recognize Nature's thinking, you can imitate it in detail. This is how we become co-creators—learn what is behind the known electromagnetic universe.
>
> Every living creature, every physical form in the act of bringing forth its visible form out of its archetype idea, passes through the swirling vortex motion in order to manifest. How could we have missed this universal machine?[60]

New energy researchers such as William Baumgartner state that their own efforts "stand on the shoulders of geniuses" such as Schauberger. In the next chapter we meet three other influential pioneers.

8

Power from the Universe

Of even greater importance to the world . . . is the production of unlimited quantities of free hydrogen . . . transmuted from the atmosphere.

—*Dr. Walter Russell*[61]

The children of the future have come of age and are ready to grasp . . . the reality of life-energy.

—*David Boadella*[62]

One energy pioneer who lived before either Viktor Schauberger or Nikola Tesla—John Ernst Worrell Keely (1827–1898) of Philadelphia—was perhaps the farthest ahead of his own time.

John Keely: musically inspired breakthroughs

As a musician and carpenter, Keely worked with sound and other forms of vibration to set machines into motion. He reportedly performed feats that twentieth century science is still unable to do.

For example, according to one of the stories discovered by Keely researcher Dale Pond[63] of La Junta, Colorado, an apprentice spent six months with Keely learning how to build a motor.

"Are you ready to run it?" Keely asked after the final adjustments were made. "Then go ahead and turn it on." The apprentice flipped the switch, but nothing happened. Keely walked over and put his hand on the fellow's shoulder, and the motor started.

A motor built to respond to a specific individual's touch? That was only one of the accomplishments attributed to Keely by writers of his day. According to historical documents, he performed other incredible feats:

- He built a machine that tunneled through rock by an unusual process that he called "dissociating" the stone. His invention seemed to dissolve the rock as fast as he could move his device forward.

- He freed the latent energy within water in a manner similar to work being done by today's researchers in sonoluminescence, in which tiny bubbles are created in water by sound waves and energy is released when the bubbles implode. Observers saw an engine run on the energy freed by this device, which Keely called his Liberator.

Keely discovered these and other effects by experimenting. His musical background allowed him to craft machines similar to musical instruments. As in the building of a violin, the machines were built to respond "sympathetically" to tones in harmony, not conflict. But Keely's machines depended a great deal on what he called the "vibration tones" of the builder—such as the person's breathing, brainwave rhythms and intention. It was as if a violin could only be played by the person who made it. Therefore, engineering his machines was not a simple matter of tuning mechanical resonators; they were much more sensitive than standard machines.

Keely reached an advanced understanding of the science of vibrations and what we now call *quantum physics.* His level of knowledge exceeded our own modern accomplishments, Pond says. At first only Keely could operate his machines; they were keyed to his personal vibratory signature. After three years of work he developed a way others could also operate his machinery.

According to Pond and other new energy historians, Keely discovered more than forty of what he called fundamental laws of nature. Among other accomplishments attributed to Keely are the creation and control of frequencies at an extremely high range, as well as work in the fields of acoustical motors, ultrasonics, and control of extreme pressures and vacuums. His major work was with *sympathetic vibrations* on the atomic level—what we now call *quantum entanglement*. Sometimes it is hard to remember that this work was done in the nineteenth century.

Why isn't Keely's work well known today, and why isn't it studied in the scientific community? Part of the reason seems to be that he was so far ahead of the science of his day that his work was simply dismissed by most (but not all) scientists. His paradigm was so far beyond the worldview of the time that very few could make the intellectual leap required to grasp it, Pond says.

Another reason that Keely disappeared from history is that, while he claimed to have written thousands of pages explaining his work, his writings have yet to be found.

Dale Pond, Dawn Stranges with Keely-type invention.

Gentler science

Conventional physics often takes the sledgehammer approach—slam an atom with energy, break it, and see what is in it. This is the opposite of Keely's approach. He saw the atom as being like an orchestra, an assortment of vibrating parts producing various tones. This is more commonly accepted by modern physicists, but it was a revolutionary idea in Keely's day.

What really set Keely apart was his ability to act as the conductor of this atomic orchestra, his seeming ability to get the atom to do his bidding. Pond says that Keely discovered, among other procedures, how to bring two vibrations together so that they formed a third, different vibration. He uses the example of an opera singer shattering a wineglass with not only perfect pitch but also the sustained volume of her voice, which represents the conventional sledgehammer approach to breaking up matter. In contrast, Keely's approach would be to melt (loosen the atomic bonds) and reshape the wineglass by adding a tiny amount of acoustic energy at just the right pitch.

Debunkers thrive on the Keely story, amusing their readers with the unusual names of his inventions—Disintegrator, Sympathetic Transmitter, Vibratory Accumulator, Tubular Resonator—and his extraordinary claims. There were also allegations of fraud. Newspapers of Keely's day considered the case closed when an investigator found a large metal sphere buried under the floor of his laboratory and slender pipes running throughout the walls after Keely's death. They declared that Keely had been using compressed air to power his experiments.

Researchers with some understanding of what he was doing, on the other hand, explain that was not the case. They say the sphere was not a hidden trick but a part of an experiment that was later stored under the floor. They also say that Keely used the pipes—too small in diameter to supply compressed air with any force—in advanced experimentation. The pipes carried what he called his *Etheric Vapor* around his laboratory to various experiments, much as modern mechanics conduct compressed air to machines in their shops today.

Keely began experimenting with vibrations and energy in the early 1870s. By 1874, he had gained some mastery over this force, which he called "liberated ether," but he had run out of money.

Mystery of the writings

Despite Keely's business problems and the subsequent negative newspaper reports about him in his era, Dale Pond says that modern science is slowly vindicating his work.

Now that he is being taken more seriously, the Keely mystery deepens. What happened to the bulk of his writings and the hundreds of machines he made? No one is sure. But researchers such as Pond are rebuilding Keely's machines and continuing his experiments.

Walter Russell and the invisible geometry of space

Frontier scientists who want to tap into the energy of universal space in order to power useful inventions learn about invisible aspects of the universe. They find it operates along preferred paths and shapes or geometries.

The job of these scientists is similar to that of other creative people:

- A designer needs to know a fabric's characteristics to create something special with it.
- A book printer needs to know which way a roll or sheet of paper's grain is oriented.
- A jeweler must understand the crystalline structure of a gem before shaping it.
- A chemist learns about the reactions that result from various mixtures of compounds.
- Weather forecasters study the interaction of air masses of various temperature and moisture content and learn to predict their behavior.

Dr. Walter Russell (1871–1963) studied how matter and energy in the universe continuously cycle in a spiraling two-way exchange like a pump that perpetually compresses and then expands a substance. Russell learned that the universe consists of electric energy, and that nature multiplies power by concentrating this electricity—or background energy—until it forms atomic matter that later assembles into large aggregations such as a star or a planet.

Walter Russell, renaissance man

Around the start of the twentieth century, Russell was called the most versatile man in America, a Leonardo DaVinci for modern times. As an acclaimed artist, architect, musician and poet with a Van Dyke beard, Russell appeared unlikely to also be a scientist. He sculpted masterpieces including a statue called the Four Freedoms, commissioned by President Franklin D. Roosevelt.

The first impression one author[64] had of Russell was "a light in his eyes that showed that he was capable of great inspirations—that he lived close to the great unseen powers of the universe."

Russell successfully forecast six atomic elements that previously were not known.[65]

After several science laboratories isolated the elements which Walter Russell had predicted would be found, in 1941 the American Academy of Sciences gave him a doctorate degree.

Walter Russell said that elements can be induced to change to a different element by rather simple means that establish conditions favorable to one over the other. He believed this discovery could be used to establish a pollution-free hydrogen economy based on abundance instead of scarcity. A cost-effective process for such elemental transformation, if it is developed and accepted, it could also help in cleaning up radioactive waste and other toxic substances.

In the 1990s, three Colorado researchers performed in-depth laboratory research to test Russell's theories. Electrical engineer Toby Grotz, Dr. Timothy Binder and analytical chemist Ronald Kovacs repeated a Russell experiment from 1927—verified in that year by Westinghouse Laboratories—that demonstrated the changing of one element to another by a low-cost method. The unorthodox experiment did not come to the attention of mainstream scientists, however, and the work progressed no further after the trio moved on to separate projects.

Military interest

Russell also built a device, which he named the Russell Optical Dynamo Generator,[66] that he claimed had captured the background energy found everywhere. The original blueprints of this device were found in a Colorado basement. The owner was an associate of a general with the North American Air Defense Command (NORAD), a defense agency responsible for protecting North America from missile attack. Russell had worked on his device with both NORAD and scientists from the Raytheon Corporation, now a major contractor to the Pentagon and U.S. intelligence agencies.

In 1959, officers from NORAD in Colorado Springs visited Russell and his wife and research assistant, Lao, at their home in Virginia. They agreed that the Russells would report on their findings. On September 10, 1961, the couple reported that the Russell generator had worked, and that the president of the United States could announce to the world that a new, safe power source was available.

However, the Russells' conviction that they had demonstrated a way to convert background energy into electric power failed to capture the interest of anyone except NORAD, and there is no public record of what NORAD did with it. At the time, conventional science termed the discovery "nonscientific," and the public never heard mention of it.

The Russells founded the University of Science and Philosophy in Waynesboro, Virginia. Trying to explain cosmic laws to the established scientists, Russell published a book titled *The Universal One* and then worked another twenty years on his explanations and published *The Secret of Light*. Some of today's leading frontier scientists recommend reading Russell's writings to better understand electricity and the invisible geometry of the motion of space. In Russell's science, that motion causes gravity, magnetism and electricity.

Frontier scientists who agree with Russell say entropy is not the only game in the universe; we also have "negentropy" or "syntropy" where systems are running uphill to higher energy states. Not only do natural systems degenerate to disorder and chaos, but they also rise out of chaos to order. To build new energy systems based on what Russell called the whole power of the universe, would be to integrate the interchange between two opposite conditions. Power can be tapped into at both ends. Humankind has predominantly tapped it only on the expanding end—things coming apart. The zero point energy being discussed in leading-edge physics means harvesting energy from the contracting spiraling of universal space, instead of from the expanding degenerating spiraling in which matter is disintegrated.[67]

Russell advised people to model nature's generous ways—not just in technology that works for everybody but also in our social systems.

Wilhelm Reich and the orgone motor

Like Walter Russell, the inventor of another new-energy device was a highly educated man. Wilhelm Reich (1897–1957) was an Austrian—later an American—scientist and outspoken innovator in fields ranging from psychiatry to biology. All of his work led toward one unifying discovery—a pulsating, life-force energy found everywhere, in varying degrees. Reich named this energy "orgone," because he first discovered it in living organisms.

In 1948, the famed educator A.S. Neill of England saw a small motor running in a workshop in Maine. It was connected only to an "orgone accumulator," without any other source of power. His friend Wilhelm Reich was the proud inventor of the setup. "The power of the future," Neill said, yet Reich made no attempt to progress the technology.

Why didn't Reich, a prolific innovator, continue this line of research? "My job is discovery, and I leave it to to others to carry out the results," he said in a letter to Neill.

Overlooked free energy

Although Reich first discovered orgone while doing research in psychiatry and biology, he found that it could be used as a motor force. In 1947, he bought a Geiger counter to detect cosmic radiation, since he thought orgone might have properties similar to the cosmic rays that constantly stream into our atmosphere from space. When he put the counter inside an orgone accumulator—a box designed to capture and concentrate orgone—the counter registered the normal background radiation by clicking at a normal rate, about thirty counts per minute. Reich then turned to other projects and stored the Geiger counter next to a miniature orgone accumulator.

A few months later, he picked up the Geiger counter and tried it again, and found it was clicking along at an amazing 6,000 counts per minute. After performing some tests, Reich decided the Geiger counter had become saturated with orgone energy. A year later, he found that vacuum tubes—the kinds of tubes that used to be found in television sets—were also affected by long soaking in the concentrated orgone environment. These tubes showed the strong effects of orgone by giving off a strong violet-blue light. This led Reich to try using the tubes to run a motor, a successful feat witnessed by five members of his research staff.

Based on these and other experiments, Reich decided that static electricity and orgone are related. Static electricity is electricity at rest, as opposed to dynamic electricity, which flows as current in a circuit. Static electricity causes hair to stick to your comb in winter and dynamic electricity flows through the wires in your home.

Reich thought that orgone was the one primary energy, similar to static electricity in that it permeates large areas without irritating the living beings in the vicinity. Dynamic electricity, by contrast, is a coarser form of energy that irritates living beings.

Such a conclusion—that orgone and static electricity are related—could provide insights in some areas of new-energy investigation, such as electrostatic motors powered by the energy in the space surrounding them.

Reich's books are burned

One reason that Reich did not pursue the development of the orgone motor was his ongoing fight with the Food and Drug Administration, which took up his time and energy for years. In 1954, the FDA, in an attempt to suppress the use of the orgone accumulator in physical therapy, ordered Reich's hardcover books banned from circulation. His soft-cover books, including all his periodicals, were burned by government employees. Anything that mentioned the word "orgone" literally went to blazes. For refusing to obey an injunction against publishing his material, Reich was sentenced to two years in jail. He died in prison in 1957.

In his long and prolific career, Reich voiced many unusual observations, especially when under stress toward the end of his life, such as his claim that his "cloud-buster" machine could affect UFOs. Therefore, he has been another easy target for skeptics who focus on an innovator's amusing eccentricities rather than on his accomplishments. They also focus on untruths. Misinformation about Reich is repeated so often that popular writers refer to sex experiments in connection with orgone accumulators as if factual.

Continued interest

A few new-energy researchers have tried to build the orgone motor, but found that Reich left inadequate notes. One inventor was so intent on getting information that he broke into the Wilhelm Reich Museum in Rangeley, Maine. He was imprisoned for burglary, and the information was returned to the museum. Other researchers are looking into orgone's medical effects.

Today a small number of scientists with an understanding of orgone energy are carrying on with experiments that prove there is a real effect in the orgone accumulator that cannot be explained by conventional science.

In Canada, Paulo Correa, PhD and Alexandra Correa[68] followed the trail of Reich's knowledge of orgone energy into the realm of energy-conversion devices. Our friend the late Eugene Mallove, PhD, witnessed successful demonstrations at the Correas' laboratory.

Schauberger, Keely, Russell and Reich provide intriguing examples of those researchers whose curiosity leads them to explore and harness the mysterious natural energies found in water, sound, air, and space itself. The pioneers in the next chapter worked on the frontiers of knowledge about magnetism.

9

Magnetic Pioneers

There are secrets and mysteries surrounding magnets and collapsing field energies, and only after exhaustive studies . . . do these mysteries unravel themselves and emerge in their glory.

—*Robert Adams, PhD*

The magnet is a window to the free space energy of the Universe.

—*Bruce DePalma, PhD*

Hundreds of people witnessed dozens of demonstrations that Mr. Gray gave in his laboratory.

—*Dr. Peter Lindemann*[69]

The cover story in the spring 1980 issue of Popular Mechanics magazine had a dramatic title, "Amazing Magnet-Powered Motor."[70] Its writer, a former research scientist, had crossed the continent to Blacksburg, Virginia, to meet the inventor of the motor. Within two days, the formerly skeptical journalist said, he had become a believer; Howard Johnson demonstrated working devices that created motion using only permanent magnets. Johnson had a new patent[71] on his motor and said its power source was only the energy contained in the atoms of permanent magnets. The Johnson linear device (movement in a straight line rather than rotating) was not practical for powering appliances but it did show that a device unconnected from the power supply could be powered by magnets.

Johnson was one of a long line of inventors, going all the way back to Peter Peregrinus in the year 1279, who researched how to build magnetic motors and generators that run without an external power supply. Conventional teachings say that no work can be performed solely by use of permanent magnets, but unconventional findings indicate the textbooks will have to be rewritten.

Breakthrough with no moving parts

The late Floyd "Sparky" Sweet created a magnetic solid-state energy generator with no moving parts. Although he was a magnetics specialist with a distinguished industrial career, for complex reasons he did not develop his device into a commercially viable product. However, Sweet's story is important for three reasons. First, creditable witnesses saw his invention convert the invisible energy of space into useable amounts of electric power—without fuel, batteries, or connection to an outlet. Second, he was subjected to the same kinds of harassment that some other inventors we met had to face, including threats on his life. Third, and most important. Sweet's research has inspired the work of other new-energy inventors, some of whom may well produce a useful stationary-magnet device.

Floyd Sweet and magnets

Floyd Sweet (1912–1995) grew up in an era when radios were home-built crystal sets. At the age of nine, his intense interest in how things work was directed into building and disassembling radios and other electrical apparatus.

When Sweet was eighteen, a family friend helped him find work at the nearby General Electric plant while he went to college. He got the nickname "Sparky" after he misconnected some wires one day and an instrument exploded in a spectacular spray of sparks. Despite this incident, his employers were pleased with his work—especially his intuitive gift for coming up with answers to electrical problems.

Sweet worked in GE's Schenectady, New York, research and development center from 1957 to 1962, a dream job in which he could use a well-equipped laboratory to follow his hunches on intriguing magnetics projects. That line of research fascinated him. He augmented his technical credentials in 1969 with a master's degree from the Massachusetts Institute of Technology.

By the mid-1970s, Sweet and his wife, Rose, had moved to the Los Angeles area to enjoy semiretirement. Besides serving as one of GE's preferred consultants, Sweet designed electric equipment for other customers.

Floyd Sweet had a passion for magnetism, and for the concept that the entire universe is permeated with a magnetic field. Once he fully retired in the early 1980s, he would have happily spent many hours each day building a device that could tap into the energy of that magnetic field.

But Rose fell ill, and was an invalid for the last seven years of her life. This demanded Floyd's attention and forced him to dip into their savings. He also had to cope with his own ill health. Despite these problems, he worked on his device when not preparing meals and tending to his wife's needs.

Shaking the magnetic field

For decades, new-energy researchers talked about the possibility of treating a magnet so that its magnetic field would continuously shake or vibrate. On rare occasions, Sweet saw this effect, called self-oscillation, occur in electric transformers. He felt it could be coaxed into doing something useful, such as producing energy.

Sweet thought that if he could find the precise way to shake or disturb a magnet's force field, the field would continue to shake by itself. It would be similar to striking a bell and having the bell keep on ringing.

Sweet—who said his ideas came to him in dreams—turned for inspiration to his expertise in magnets. He knew magnets could be used to produce electricity, and wanted to see if he could get power out of a magnet by something other than the standard induction process.

Induction involves either moving a magnet past a wire coil—a coil of conductive wire, such as copper—or moving a coil through the field of a magnet. This changing magnetic field causes an electric current to flow in the copper wire.

What Sweet wanted to do was to keep the magnet still and just shake its magnetic field. This shaking, in turn, would create an electric current. One new-energy researcher compares self-oscillation to a leaf on a tree waving in a gentle breeze. While the breeze itself isn't moving back and forth, it sets the leaf into that kind of motion. Sweet thought that if cosmic energy could be captured to serve as the breeze, then the magnetic field would serve as the leaf. Sweet would just have to supply a small amount of energy to set the magnetic field in motion, and space energy would keep it moving.

By 1985, he had come up with a set of specially conditioned magnets, wound with wires. To test his device, Sweet discharged a current into the wire coil around the magnet. As a result, the coil disturbed the magnet's field. It was as if Sweet had snapped the magnet's field out of position to set it in motion. Sweet then connected a twelve-volt lightbulb—the size used in flashlights—to the coil. If the device was producing electricity, the bulb would light.

The results were more than Sweet expected. A surge of power came out of the coil and there was a bright flash from the bulb, which had received so much power that it melted. Years later, Sweet remembered that Rose had seen the flash and called out, "What did you blow up now?"

The inventor was baffled by the dazzling flash of light—why so much energy? He returned to his workbench to make further models. Needing a theory to explain his startling discovery, he remembered hearing about Thomas Bearden, retired Army officer and nuclear physicist, and John Bedini, an electronics expert, on a local radio show. Sweet called Bedini, who arranged for Bearden to visit Sweet.

Bearden witnessed the curious device produce an output of nearly six watts of electric power with only a tiny fraction of a watt going into the machine. Bearden ran tests to his heart's content, and was delighted to see a little unit embodying the unorthodox concepts that he had written about over the years, the concepts behind energy from the "vacuum" of space. He named Sweet's assembly of magnets and wire coils the Vacuum Triode Amplifier (VTA). Bearden decided that the device was serving as a gate through which energy from space was being herded into an electric circuit.

The most amazing aspect of Sweet's device was that it put out so much more power than it took in. How much more? In a 1988 model Sweet found that 330 microwatts, 330 one-millionths of a watt, of input power made it possible for the VTA's wire coils to put out more than 500 watts of usable energy, or about one and a half *million* times the input power.[72]

Special effects and difficult development

The VTA turned out to have some very odd effects, but Bearden's research background prepared him for that. So in 1987, Bearden asked Sweet to perform an antigravity experiment. Bearden calculated that the six-pound machine would levitate when about 1,500 watts of power were drawn out of it, but the magnets might explode at about the same power level. He warned Sweet to limit the output to no more than 1,000 watts. A VTA would be placed on a scale so that its weight could be carefully monitored while it was hooked up to a box of lightbulb sockets. Screwing bulbs into the sockets would draw off the power.

About a week later Sweet excitedly read off the results over the phone to Bearden, who was home in Alabama, as Sweet screwed in ten 100-watt bulbs, one at a time. The device gradually lost weight until it was down to 90 percent of its original weight. For safety reasons Sweet and Bearden stopped the experiment before the device could begin to hover or fly.

Why did the VTA lose weight? According to Bearden's theory, gravity becomes a pushing force rather than a pulling force under certain conditions. Bearden also says that space energy has a pressure, referred to as energy density. If the pressure above an object is decreased while the pressure under the object is increased, the object will be drawn upwards. The VTA may have changed the energy density by drawing on cosmic energy.

Another unusual effect of Sweet's VTA was the fact that it produced cold, instead of the heat usually generated by electric equipment. The inside of the VTA was as much as twenty degrees cooler than the surrounding air. The greater the load put on the device, the cooler it became. When VTA wires were accidentally shorted out, they flashed with a brilliant burst of light, and were found to be covered with frost. One time, a brief contact with the equipment froze some of Sweet's flesh, causing him pain for about two weeks afterward.

Sweet discovered other interesting effects. But development of the

VTA was slowed by trouble with materials and processes, and by financial entanglements. Sweet had to find magnets that could hold the self-oscillation effect. That required magnets with force fields that didn't vary much across the face of the magnet. Also, standard mathematical calculations didn't work with the VTA, so Sweet produced a math theory for the VTA—an engineering design model.[73]

As with first models of any new technology, the VTAs he built were very unreliable. For example, at times their output went down at night and picked up again during the day. Frontier scientists who do experiments with the orgone energy described by Wilhelm Reich say that the density or strength of the background energy in the atmosphere has natural cycles and pulses like a heartbeat. Could the VTAhave tapped into that more primal level of energy? If so, then variations in its operation could be explained if not controlled. Sometimes, the devices just plain stopped working for no apparent reason. But when the VTA worked, the power it put out for its size was unprecedented.

Challenging the laws of physics

Bearden contributed to the theory that explained Sweet's invention.[74] Much of the theory that Bearden used to explain how the VTA worked came from advances in a specialized study of light[75] used by laser scientists and weapons researchers. Using information from this field, Bearden said that the VTA was able to amplify the space energy it took in.

The science establishment requires that an invention be explained by accepted laws of physics, and so much output from so little input *seems to* violate those laws. However, Sweet and Bearden recognized that these laws apply to ordinary, or closed, systems—systems in which you cannot get more energy out than what you put in. Because the VTA allowed energy to flow in from the vacuum of space, it was not operating in a closed system, but in an open one.

A device operating in the flow of space energy is like a windmill operating in the wind. Both receive excess energy from an outside source. But since neither operates in a closed system, neither violates the laws of physics.

In 1991, a paper by Sweet and Bearden was read at a formal gathering of conventional engineers and physicists in Boston.[76] Neither Bearden nor Sweet were able to attend—Bearden was called away on business, and Sweet was recuperating from heart surgery. Walter Rosenthal went instead. (The late Walter Rosenthal of California was a test engineer who helped many struggling inventors test their devices.) The paper he presented said that the VTA had the signs of being a true negentropy device, or a device that was able to turn random background cosmic energy into usable electricity.

Order out of chaos?

"*Entropy*" is a word for the commonly observed tendency of systems to degrade to lower energy and greater disorder. Typical human experience suggests that fragile objects placed up high tend to fall and randomly scatter pieces when they break. A crystal goblet on a counter top has more potential energy and more order than the same glass that has fallen to the kitchen floor and fragmented into hundreds of pieces. A well-maintained home left untended for months soon shows sign of disrepair.

We are familiar with such things. We are not familiar with the reverse process of fragments of glass spontaneously forming themselves into a goblet and "falling up" onto the countertop perfectly upright, or a house with peeling paint and leaves in the gutters spontaneously returning to showcase form—which if they occurred would be examples of "*negentropy*", or a tendency toward greater order and higher energy.

Although the language they used was quite technical, what Sweet and Bearden basically said is that the VTA was able to take random energy fluctuations from space and organize and harmonize them to an apparent higher energy level. As a result, the device demonstrated a build up of energy in its output rather than the energy dissipation typically observed with other systems. Most conventional physicists would flatly and dismissively declare this to be impossible without any investigation.

After Bearden's paper was read, Rosenthal stood up and startled the audience of skeptical engineers: "I have personally seen Floyd Sweet's machine operating. It was running . . . those small motors you saw in the video. It was jump-started with a nine-volt battery. There was no other electrical input required . . . . There was no connection to the power line whatsoever." And, no, there were no moving parts.

Although most of the audience listened politely, it was too much for one engineering professor. He stalked out of the room, saying, "To present such a remark at an engineering conference is the height of irresponsibility! It violates virtually every conceivable concept known to engineers."

Threats

Could activity at the Sweet home have been secretly watched by strangers? Sweet told the story of a time in the late 1980s when a man accosted him as Sweet was leaving a supermarket. Sweet remembered the man's expensive-looking shoes, and the fact that he was immaculately dressed. But in the stress of the moment, Sweet couldn't focus on much else.

What made the inventor nervous was the photograph that the man held, a photograph showing Sweet at work on his tabletop-model VTA—in the supposed privacy of Sweet's own home. In what Sweet said was a remarkably clear photo, he was sitting in the dining room of the second story apartment where he lived with Rose.

"He walked me all the way to my building, telling me what would happen to me if I didn't stop my research," Sweet recalled.[77] "How they took that picture through my window, I'll never know."

As Sweet remembered it, the man claimed to be connected with a conglomerate that did not want the VTA to come onto the market at that time. He told Sweet, "It is not beyond possibilities to take you out of the way."

Sweet said that afterward he called the FBI in Los Angeles. He believed that two agents staked out his house for a couple of weeks, but nothing came of it.

Around the time of the photo incident, Sweet told us, he was getting telephone calls and death threats from strangers, and calls at all hours. "The police put a tap on my line and over a six-month period, over 480 calls came in from all parts of the United States. But they were from pay stations." Thus, the police could never find the callers.

Early in the VTA's development, someone broke into Sweet's apartment and stole his notes. He then began to code his notes.

Sweet temporarily stopped work on his invention, out of concern for his ill wife. "They must have known I stopped; they didn't torment me any more."

Floyd Sweet suffered a fatal heart attack at the age of eighty-three, on July 5, 1995. A couple of weeks before his death Sweet said that the automotive industry was testing his power unit for use in cars, and that they had a unit running for 5,000 hours. He said he was dealing with people at General Motors, but no one has been able to confirm Sweet's claims.

After his death the VTA was bogged down in legal problems and the confusion surrounding Sweet's business affairs. But Tom Bearden, who put much of his own time and money into the project, hopes that the

Sweet's legacy is well-known on discussion groups that list "free energy" as a topic. Experimenters seek details of how the device was built. On his personal website[78] Thomas Bearden is releasing information and sells a DVD titled "Floyd Sweet's Secrets."

When this fledgling space-energy science reaches maturity, what could the VTA do for the lives of the rest of us? Bearden has for years been speculating that the new physics will change our lives in undreamed-of ways:

"By mastering, controlling, and gating the vast, incredible energy of the seething vacuum [of space], we can power our automobiles, flying machines, and technology inexhaustibly. Further, it can be done absolutely cleanly; there are no noxious chemical pollutants."

Despite the difficulties that Sweet ran into in his attempts to perfect his invention, he helped science take a leap into the future. It perhaps could have leaped further if he had cooperated more freely with other researchers in the last decade of his life, and if he had been tidier in his business dealings. But Sparky Sweet deserves praise for charting a new course.

Graham Gunderson (see Chapter One) has taken on another challenge in addition to developing his own Genie invention. He is applying his knowledge of oscillating magnetic fields in researching Sweet's discoveries, with the intention of replicating Sweet's experiments. Gunderson's goal is to get the technology into the marketplace for the benefit of humankind.

Gary's magnetic motor

Other energy pioneers also discovered the power of magnetism and magnets in motion.

In the nineteenth century in Pennsylvania, an inventor named Wesley W. Gary created a fuel-free magnetic motor. He made a discovery that had escaped the notice of scientists and electricians—the existence of a "neutral line" in the magnetic field of magnets. [The neutral line is where one polarity ceases, and beyond it another polarity begins.] As a result of that discovery Gary patiently developed magnetic devices that worked without batteries. A popular magazine said in 1879, "The Gary Magnetic Motor, the result of Mr. Gary's long years of study (is) as simple contrivance which furnishes its own power, coming dangerously near to that awful bugbear, perpetual motion." (Harper's New Monthly Magazine, March 1879, pp. 601–605) . Gary patented his improvement, and Harvard and

MIT professors confirmed that he had done what he claimed, but textbooks continued to deny that such inventions are possible.[79] Today, with information on the Internet[80] a few researchers are rediscovering the Gary motor, along with Walter Russell's insights on the magnetic vortex flow that creates the neutral area between magnetic polarities.

Alan Francoeur of Canada gives Wesley Gary credit for insights that went into Francoeur's "interference disk generator". Again it took years of study and months of patient work to adjust delicate balances in order to get the device working. We believe it's time for a task force to propel these discoveries into the stage of commercial development.

Coler's magnetic generator

Little is known about German Navy captain Hans Coler outside of some technical reports, but what he learned could have changed history. In 1925 he invented a small magnetic generator without moving parts and showed a ten-watt model to a professor in Berlin. Despite the professor's endorsement, the German government refused to patent the device on the grounds that it was a perpetual motion machine.

With financial support from the German admiralty ten years later, he produced 6,000 watts from a magnetic generator. "These fundamental researches have made the first real and large breach in the citadel of present scientific belief," Coler wrote.[81] He was said to have powered his house with a seven-kilowatt "Stromerzeuger" version.[82] Near the end of World War II, an Allied bomb destroyed his home and laboratory. Coler survived and later cooperated with British Intelligence. A British scientist and a Norwegian Army scientist[83] took replacements of the destroyed equipment's parts to Germany and watched Coler wind coils around the magnets, assemble them in a loop, connect wires and power light bulbs. Testimony from German professors confirmed that the no-moving-parts "Magnetstromapparat" (Magnet Current Flow Apparatus) did generate power in a mysterious way. The two scientists' report said one model of the device worked for three months while locked in a room in the Norwegian embassy in Berlin.[84]

A British secret report from 1946 on Coler's work was declassified in 1979.[85]

British magnetics expert Harold Aspden, PhD learned about the Coler mystery in 1988 while beginning to see how energy could be tapped from the aether (background energy) by exciting magnetic cores in an electric motor in a certain way. Aspden figured that the most likely reason that the Coler file was closed with the mystery unsolved was that any scientific committee would have decided that the Coler claims could not be viable "as it was contrary to longstanding experience."

Post war pioneers of magnetism

In the more recent history of magnetic research pioneers who are no longer here, the name of Ed Gray is prominent among those who invented revolutionary electromagnetic motors.

While Coler was baffling the professors, across the ocean teenaged Edwin Vincent Gray was starting out. Born in 1923 to a low-income family of six children in Washington DC, at fifteen years of age Gray left the streets and joined the United States Army. He attended its advanced engineering school until discovered to be under age and discharged. Later he learned about radar during his tour in the U.S. Navy.

The mystery of "cold electricity"

Gray was so lacking in formal education that he didn't realize for some time that his thinking was original and advanced.[86] Three well-known things about electricity fascinated him; his genius was in correlating them into a new technology. The factors had to do with capacitors[87], the fact that pulses of electricity can be sent out and brought back, and that lighting bolts are more powerful when closer to Earth.

By 1958 he had learned to perform what he saw as a splitting-out of a component of electrical current, and using this "cold electrical energy" as well as normal electrical energy. His first motor was operating by 1961 and he began preparing to put units into production.

Journalist Tom Valentine was one of the witnesses to demonstrations of Gray using a quite different form of electrical current. The inventor performed experiments that would have quickly drained a battery. If powered by a battery, the experiments would have made Gray's magnets extremely hot. They remained cool.[88] This was an advantage because in a motor, cool-running parts do not wear out as quickly as overheated ones.

The main advantage of Gray's breakthrough is the possibility of powering engines "without spitting a single speck of pollution into the earth's atmosphere."[89] He said his invention could be sized up to 1,000 horsepower ouput with a single unit, or be scaled down to toy size.[90]

Peter Lindemann explains that Gray had discovered that the discharge of a high voltage capacitor could be shocked into releasing a huge, radiant, electrostatic burst. Gray captured this energy spike in a special device. "The non-shocking, cold form of energy that came out of this 'conversion tube' powered all of his demonstrations, appliances, and motors, as well as recharged his batteries. . . . During the 1970's, based on this discovery, Mr. Gray developed an 80-horsepower electric automobile engine that kept its batteries charged continuously."

Harassment of Gray began in 1974 when the Los Angeles District Attorney confiscated all Gray's records and his working prototype. For eight months, the D.A.'s office tried to get EvGray Enterprises Inc. stockholders to testify against Gray, but none of them had a complaint.[91] However, to escape what he saw as major false charges of fraud, Gray later pleaded guilty to two minor violations of Security Exchange Commission laws and was released. His prototype was never returned. Gray died at his shop in Nevada under mysterious circumstances in 1989.

Gray's cold electricity

The implications and details of Gray's motor and "cold electricity" stir up strong interest among the network of new energy researchers today. Dr. Peter Lindemann says that until recently, only a handful of researchers have understood how to tap the "cold" form of electric power that can be produced with relatively simple equipment. Apparently cold electricity can do work just as ordinary electricity does, but without the fuel costs. However, the scientific community misunderstands it, partly because 'Radiant Energy' or cold electricity does not behave exactly like typical electricity, Lindemann notes. "This natural energy form can be gathered directly from the environment—mistakenly called "static" electricity—or extracted from ordinary electricity . . . ." The method for extracting it from electricity, however, is beyond the scope of our book.[92 93]

Other magnetic pioneers

The late Dr. Robert Adams of New Zealand invented electric motors, generators and heaters that run on permanent magnets. His colleague Lindemann reports that one such device drew 100 watts of electricity from the source, generated 100 watts to recharge the source, and produced over 140 British Thermal Units (BTUs) of heat in two minutes.

Howard R. Johnson (1919–2008), whom we mentioned at the beginning of this chapter, was an engineer who invented products for the

United States military and industry, taught college chemistry, consulted and had his own research laboratory. Johnson successfully built a self-rotating permanent magnet machine at least twice, according to his colleague Thomas Bearden, "only to have it promptly stolen in professional break-ins to his laboratory." Lack of funding necessary to afford the high precision machining of parts needed for his inventions interfered with his progress. However, Johnson left a legacy of new understandings of "asymmetric magnetic systems".[94]

And there were other magnet-power pioneers such as Bob Teal[95] who should be honored in an encyclopedia of new energy heroes.

After meeting these brave innovators of the past who stepped outside of the box of ordinary engineering, we will next travel through hidden history that mainly involved standard engineering—but also involved extraordinary political interference with a clean energy technology of vast potential.

10

Oil and Water

One month prior to his assassination, JFK committed the United States to a spectacular, sustainable vision of ocean energy. Now is the time to fulfill JFK's vision.[96]

When former United States president John F. Kennedy stood on the beach at his home in Massachusetts, he may have had the same reaction as we would—thrilling to the roar and hiss of ocean waves. Some visionaries look at the relentless currents and think "Harness this force!" Did the forceful push and tug of sea tides inspire Kennedy's vision of abundant power from the ocean?

Kennedy's own words were read into the Congressional Record in 1963.

> Man only needs to exercise his engineering ingenuity to convert the ocean's surge into a national asset. . . . I think this (Passamaquoddy Tidal Power Project) can be one of the most astonishing and beneficial enterprises undertaken by the people of the United States. . . . I understand that, measured by the customary feasibility standards, the Passamaquoddy-St. John's River project now meets that test.

John F. Kennedy also gave a passionate speech about tidal power in the Rose Garden on July 16, 1963. The dream had been with him for a long time. Earlier as a senator, he had described it at a Maine Democratic Party issues conference banquet in 1959:

> It is clear from this recent report (by International Passamoaquoddy Engineering Board) that to tap this fantastic flow of 70 billion cubic feet of water each day would be a tremendous spur to the economic growth of Maine, and New England, and the entire United States. . . .
>
> If we have leaders who are willing to look ahead—who are willing to spend money now in order to reap vast returns in the future—then we can look forward to a new supply of 550,000 kilowatts—to some one million tourists a year coming to view one of the most spectacular products of modern technology—to the attraction of innumerable new industries with growing power needs—and to the regeneration of the whole economy of Maine and Washington County in particular. . . .
>
> We are talking about a great national asset, like TVA (Tennessee

Valley Authority), the Grand Coulee Dam or the St. Lawrence Seaway.[98]

JFK's tidal power dream has been erased from mass-consumption history, but new energy researchers read about it in a smaller circulation magazine, *Infinite Energy*. The editor, Barbara DelloRusso, had spoken with an older woman, Irene Shepherd, who grew up near Maine's Passamaquoddy Bay and told about the proposals for a tidal project there.

DelloRusso learned that the first passionate backer for a Passamaquoddy project was a hydroelectric engineer Dexter Cooper. In the 1920s he surveyed 400 miles of Maine shoreline. He did it with his own money earned from Panama Canal contracts. Eventually he tested a scale model of harnessing tides in Passamaquoddy Bay where water rushed in from the Bay of Fundy. Cooper believed that the tides rising and falling 18 to 26 feet every twelve hours would bring a never-ending source of power, and wanted to create two tidal basins where the Atlantic pushes more than 70 billion cubic feet of water into two bays. That's more water than flows down the Mississippi River in two weeks. When the tides ebbed, water captured in the bays would be released to flow through turbine generators on the way back to the ocean.

The 1929 stock market crash took down that plan.

It was resurrected by president Franklin D. Roosevelt as a smaller-scale proposal. The engineer and colleagues then figured seven miles of locks and dams would take three years and $36 million to build. However, three power companies that controlled 95 per cent of Maine's electricity fought against the funding legislation. The utility companies persuaded the legislature to block the bill. The visionary engineer died without seeing his ambitious plan come to fruition.

There were no more strong champions for the project until John F. Kennedy. JFK swam against the tide with this dream. The Electric Council of New England, backed by private power companies, paid for a negative report on the tidal project.[9] A month after a flurry of criticism of the project in the media, it died when Kennedy was assassinated.

A tidal power proponent[100] a few years ago suggested that Jeane Manning phone a woman who has done prodigious research into Kennedy's dream, from Maine to Montreal to France.

Surprisingly, the researcher was a busy realtor in Miami's affluent

South Beach. Andrea Silverthorne is middle-aged, and from a long line of Republicans. An online realty photo showed a conservatively dressed, attractive dark-haired woman. Silverthorne told Manning she was uncomfortable in the role of investigator of a JFK story, and has tried unsuccessfully to unload the burden to mainstream media journalists. At the time of conversations with Manning, she had accumulated a four-foot thick pile of documents that alternately intrigued or frightened her.

She ran into the JFK story by coincidence. As a child she spent summers in America's easternmost town, Eastport, Maine. Her uncle was a long-time Republican state senator, representing Maine's Washington County. Children in the family heard him tell many stories about former president Franklin D. Roosevelt's vaunted attempt to build the Passamaquoddy Bay dam nearby. Locally the place was called 'Quoddy. But her uncle strangely never said a word connecting John F. Kennedy with 'Quoddy. Years later she wondered: was it politically taboo?

As an adult she finally did hear about JFK's dream, but only by chance. She was helping her aunt, the family historian, find an island that once belonged to their family. Using her realtor skills, Andrea discovered that the Canadian government had mistakenly seen the little island as abandoned. While Andrea tried to get the island back, in 1989, she searched for treaties or local history. To her surprise, one Maine history book referred to the fact that President Kennedy had plans for a tidal dam.

"I was flabbergasted that I didn't know, because my uncle had never said anything and it was in his back yard!"

Oil port wanted, tidal power not

Her politician uncle had talked incessantly about every other issue, especially about trying to get an oil port for supertankers into the Passamaquody region in the early 1970s.

Silverthorne asked her mother if she had heard of Kennedy's plan for tidal power. "She said it couldn't possibly be; she would know that. . . . We thought it was very very odd."

Andrea Silverthorne forgot about it for eleven years. Then she enrolled in university classes in journalism and film. One such class compared history with Hollywood's version. Andrea picked *Amadeus* to study, while a fellow student chose the film *JFK*. The instructor had strong opinions, praising film maker Oliver Stone's thoroughness. He was impressed that

Stone had paid Harvard University academics thousands of dollars to research the John F. Kennedy assassination.

Silverthorne raised her hand. "Then why didn't Harvard tell Oliver Stone about the Passamaquoddy project?"

"Wha-a-at?"

After she replied to her instructor's startled response, he wouldn't let her drop the topic. She then began serious research in North America and also called on a French contact to find out more about tidal power.

France does tidal

France has the world's first large tidal project, built between 1961 and 1966 in a gulf off Brittany. Its 240-megawatt generating station is at the mouth of the Rance River estuary. The station had already operated successfully for years when French experts admitted they wouldn't start another tidal project. Why not? Their answer was that it couldn't compete with France's new nuclear power plants.[101] [102]

However, a writer[103] investigating the decision, had quoted French engineers as saying the Rance project cost 25 per cent *less* for concrete than conventional French hydro plants of matching capacity. And it had extra benefits—a road across the dam made a shortcut for vehicles. At the time of the article, 1973, Britain was looking at the Severn estuary in western England, where tides push the water level up and down by forty-five feet and could supply a tenth of the country's electricity. Just looking, though.

The half-mile long dam at La Rance was renovated in 1997 and now generates 90 per cent of Brittany's electricity. "It ran 30 years before it needed one thing done to it . . . and La Rance is a peanut compared to what France has off its shores," Silverthorne says, citing other potential sites that remain untapped.

Politics is the reason that France and England didn't become the Saudi Arabia of tidal power, she concluded. "My researcher in France lived close to La Rance and had never heard of it." Charles de Gaulle, French president from 1958 to 1969, strangely downplayed his country's 1966 world-class achievement, Silverthorne says.

Back in the United States, the eloquence of John F. Kennedy had been silenced by then. The Russians completed a tidal power project on the White Sea in 1969, but after a regime change no subsequent projects were started.[104] Today the British, Norwegians and others are beginning to exploit the potential of their tidal power, but vast ocean resources from Eastport, Maine, to Cook Inlet, Alaska, remain untapped.

Information buried

Andrea Silverthorne planned to do a screenplay and a documentary related to the JFK story. Getting information was a challenge, however. When she returned to look for the one book that mentioned the Passamaquoddy dam, she could not find it.

Silverthorne said she did discover a book about Maine's history by an academic at a university in Maine. The book covered Roosevelt's involvement with Maine politics yet never mentioned Kennedy.

"Which could mean that either: a) the author omitted Kennedy's role on purpose, or b) the knowledge had been sequestered so well that even a brilliant PhD didn't know Kennedy tried to do this dam! Either A or B!" Silverthorne's voice over the telephone sounded indignant. "How can you be a brilliant PhD and write a book on Maine history, mention Roosevelt and not mention Kennedy?"

Follow the paper trail—if you can find it

Silverthorne said the story of Kennedy's tidal plans was buried. "I found out from a researcher in the Smithsonian that Department of Interior records from the Kennedy era had not been turned over to the archives, when it's the law that the National Archives gets these things. All the classified documents, State Department, all those things that are *really* classified have been turned over. Department of Interior records haven't? Give me a break!"

Even old newspaper articles about it were hard to come by. Eventually Andrea Silverthorne found them, but not necessarily in her country.

Harnessing ocean energy at Passamaquoddy was to have been a joint Canada/United States project and the two countries were to share the benefits. Canadians involved in the proposal had the habit of reading Maine newspapers, and sent some clippings to Canadian officials. After Silverthorne began corresponding with the tidal power company Blue Energy, ocean energy researcher Michael Maser scoured archives in Ottawa and sent a two-inch-thick pile of articles to her.

She also ordered a stack of papers from the university-archived Carl Albert Center Library in the midwestern United States. The packet arrived just before she left to do research at the John F. Kennedy library, so she set it aside as she headed out the door to Boston.

On September 11, 2001, she was in the Boston subway on the way to the JFK Library and Museum to study their picture archives when she heard about the destruction of the World Trade Center. She lost the zest for her project.

Space race trumped tidal dream

Silverthorne dropped the research for months. She eventually returned to Boston and looked for mention of Kennedy's tidal power vision, but found nothing. "You go into the Kennedy Library and it's the Space Center this and the Space Center that. Going into space was somebody else's dream. He supported it, but it was the scientists who pushed for it. Kennedy started pushing for 'Quoddy' in 1952! He had the dream of Passamaquoddy."

One day in her office her attention was idly caught by the materials previously ordered from the Carl Albert library. The file's contents amazed her and she was inspired to do additional research in Montreal and France.

Power struggle

The library materials gave her new insights—a nationalization angle. Articles on an intense public-versus-private power debate ran from 1960 to 1963, the same time as Kennedy's presidency.

The story led back to Depression times before she was born—to the beginnings of the ongoing battle between private power corporations and public utilities. The public sector included Rural Electrification Associations (REA). Private companies feared that rural cooperatives would lead to nationalization of electrical power generation and distribution. Companies were still railing against nationalization in mid-century, when JFK advocated a tidal project that the private companies labeled "communistic." Private-power advocates did agree with Kennedy that in 20 years America would need a new source of electricity, Silverthorne says. So what did Wall Street do? The financial sector invested in a bureaucracy, Hydro Quebec. That Canadian utility company represented nationalization, but the financial sector apparently preferred it to doing Passamaquoddy.

"You see, Quebec wasn't doing anything that threatened oil. It was a certain amount (of electricity generated), that they needed to fill the

United States' needs. It was *hydro* electric. Everybody was used to dams. It wasn't something new. They chose hydro because of the overwhelming swiftness with which tidal power could eliminate oil."

Silverthorne's sympathies were with the private sector, but not with the choice of what to do about the challenge. Her research led her to see JFK as a visionary leader who wanted to replace fossil-fuel dependency with ocean energy.

"If the public knew! Passamaquoddy was going to eliminate all the power companies in New England and the Northeast. The way I determined that was finding an article in a Canadian paper that compared 'Quoddy to La Rance, which is only one (tidal) pool, not two, and they just said La Rance was one-quarter of the 'Quoddy output, so I found the number of people it was servicing in Brittany and I found the number of people in 1960 in Maine and New England. So *one dam* took out New England. All of New England. So two, three, four dams. . . . Okay? That's what it was all about."

Passamaquoddy alone would have put out four times more electricity than La Rance. "And all the dollars invested in nuclear power would have gone down the drain if 'Quoddy had been implemented."

Through a real estate customer, in 2003 Silverthorne met an editor of a major newspaper and told the editor about her project. The editor immediately showed interest, so Silverthorne began emailing the story in segments. Regardless of the editor's interest, reportedly the editor did not think the story could go out to the American public.

Jeane suggested that Andrea Silverthorne write a book, but she rejected the idea. "I know that the story is not going to come to light until people of power let it come to light! Because it's been quashed for forty years."

Oil and water

Silverthorne likened today's world to the "chicken" race in the film *Rebel without a Cause*—"this guy in a car racing toward the cliff and his jacket gets caught in the door handle so he gets his arm hung up and he can't get the door open and he can't get out in time and he goes over the cliff. That's this world and its allegiance to oil. We're hung up on the handle of oil. And we're hard pressed to get out in time."

In 2006 Andrea Silverthorne was on the Miami Sun Post's list of The Power Women of Real Estate. She apparently had not written the book. As this book goes to its editor, however, we found a long essay by Silverthorne online[105] copyrighted 2000–2007. Climate concerns had inspired her renewed effort to share her research and to spark interest in her screenplay proposal.

New wave of interest

With rising awareness of energy issues, we believe some political tides are changing and countries are embracing various ocean energy projects. Co-author Jeane Manning has written articles about the Davis hydro-turbine, an invention that holds promise for low-impact tidal power. A company called Blue Energy, formerly named Nova, has a turbine that uses the movement of water to power electrical generation. Instead of a tidal dam that could interfere with marine life, Blue Energy engineers designed a tidal fence. The company aims to string underwater concrete caissons (chambers) on the ocean floor across places where the tides run fast. The caissons would house the hydroturbines invented by the late Barry Davis, a distinguished aerospace engineer. You can think of the turbines as the ocean equivalent of windmills that spin in either direction to tap the back-and-forth motion of tides. The tidal fence of separate turbines would allow fish to swim and boats to sail through it, unlike an ecologically-damaging tidal power project of the past.

An infrastructure can be built above the separated hydroturbines to connect the tidal fence and carry electric power lines. A highway could run along the top, connecting a mainland with an island or connecting two islands. A spokesperson for Blue Energy said it could mean profits from a toll bridge *and* electricity.

Although we would like to see it built, such a project is not at the top of our new energy list because of its megaproject scale. However, if the Davis turbine is manufactured for large projects, economy of scale developments could help bring down the price of run-of-the-river turbines. Those small turbines would be a breakthrough for powering a village on a river, or even powering a home near where an ocean pushes in and out of narrow inlets.

There won't be a Sheik of Tides sitting on the world's supply of ocean energy. Asia and Africa, Europe and the Americas and other continents are rich with energy resources—by proximity to the relentless tides and currents and waves of Earth's moving waters. We believe it is time for

John F. Kennedy's long-forgotten dream of clean ocean power to finally manifest.

In Part III of *Breakthrough Power* we will meet innovators who are alive today and carrying forward various concepts that inspired the past energy revolutionaries.

PART•III

Emerging Era

Whether designing electric cars or power plants, the imaginations of the people who design and build our world are generally limited by the prevailing knowledge of how things work. Most engineers will not invent devices that use a new source of energy if they have not been taught about that source. In this section we will see the *variety* of inventions that have come from people who leap beyond the limits of today's knowledge.

First, though, we will look at one emerging worldview, a worldview that can spread from physics to the other sciences. It is in a sense an old worldview. A century ago, science was outgrowing the old idea of a motionless source of energy called the aether or ether, which was thought to fill space like a liquid that had long ago been poured into the universe. Albert Einstein pulled together a complex theory that explained how an etherless universe could work.

Today some new-energy researchers have returned to an ether based theory of the universe but it has a twist—the theory is based on an ether in motion.

Some of the scientists in this section of *Breakthrough Power* explain their potentially revolutionary discoveries in standard science terms, by putting together known concepts in new ways.

11

Breakthrough Theories

Once we realize that nature does give us something for free, it's going to change the way we do everything.

—*KEN RAUEN, scientist*[106]

The energy density of the vacuum potential is enormous, even mind-boggling.

THOMAS BEARDEN, superelectromagnetics theorist

With a zero point energy generator, cooking, heating, air conditioning, water distillation, communication, computer and lighting needs will all be sustained with or without a crisis.

THOMAS VALONE, PhD[107]

What *is* the aether or ether or what some physicists refer to as 'the zero point energy background'?

Harold Aspden, PhD is a retired British physicist who supposes that all matter including our Earth, sun and the whole universe had to be created from an omnipresent source of energy. "And so maybe there is a way of tapping a little more energy from that source." He describes it as pervading all space and being "the seat of quantum electrodynamic activity such as the transient materialization of electron-positron pairs one can read about when studying theoretical physics." Aspden has done more than read physics; he has used his expertise in the field of magnetism in actually working on energy inventions. Meanwhile the physics establishment ignores his findings. Are we in an era similar to that of Italian astronomer Galileo Galilei?[108]

If a legend involving the Leaning Tower is true, professors of the University of Pisa saw Galileo's cannon ball and lightweight musket ball fall at the same speed from the leaning tower. Academic doctrine[109] of the time ruled out that feat as impossible, so what did they tell their students back in the classrooms? "Don't believe Galileo."

New energy conference in Switzerland has displays of devices.

As long as mainstream scientists continue to believe that tabletop devices that can tap into a previously unrecognized power source are something to be avoided like a medieval plague, explanations about the power source will continue to be a confusion of viewpoints. Until the new discoveries are acknowledged and studied by the science establishment, inventors of those machines will continue to invent words to explain their machines' operation and power source, instead of using a common terminology.

Spectrum of worldviews on energy

At conferences where new-energy inventors, engineers and physicists gather to share information, they don't all listen to every speaker. Viewpoints and jargon vary more widely than the differences between the stereotypical mechanistic mainstream physicist and proponents of the concept of a life force.

At one end of the continuum of viewpoints, highly educated persons employed in corporations or universities speak in accepted engineering or physics terms. Such people rarely take the podium at new energy conferences until after retirement, because their colleagues at work would scorn such associations. "Stay far, far away from the free energy cranks."

At the opposite far end of the continuum, a lone inventor with little formal education tries to convey what he has learned in his tinkering with devices that convert a previously unrecognized source of energy into motion or electricity. He may have read enough physics to use or misuse some of its language to describe the unusual aspects of his invention, but credibility plummets when he mixes in his own coined words or uses the podium time to talk about his religious views. "Jesus gave me knowledge of free energy to save the world."

The unfortunate part of the clashing personalities and discomfort with one another's jargon is that valuable information is heard by only a small audience.

Questioning the dogma

Closer to the middle of the continuum is the viewpoint that a scientific revolution is overdue because physics has become incomprehensible and trending toward increasingly esoteric mathematics that doesn't translate into anything mechanical and useful.

Physicist and science historian Peter Graneau PhD once said, "The indoctrination of physics students, their blind faith in what they have been taught by their elders, and the career punishment of those who challenge the consensus metered out in textbooks, is common knowledge since Galileo's time.[110]

Many such observations from numerous heretical scientists with prestigious credentials raise questions about whether establishment scientists of our time know everything. If they don't, then perhaps what was considered impossible may be possible. Not impossible perpetual motion from a machine that supposedly creates its own energy, because that *is* impossible, but revolutionary ways to convert difficult-to-detect unseen energy could be explained.

It's an open system

Elsewhere on the spectrum of viewpoints is an equally practical researcher who lives in Tesla's former homeland, Croatia. Zlatko "Shad" Loncar says even self-running permanent magnet motors are impossible. It's a matter of wording. The motors always will be a *medium* for energy, Loncar explains. Magnets *receive* their energy in an open system. So the converted energy still comes from outside the device, rather than being self-generated, which takes nothing away from the tremendous breakthrough clean energy value of a practical working magnetic motor.

Windmills are open to a source of free energy that you can see as it bends nearby tree branches. You can see the brightness of sunlight reflecting on solar panels although you cannot directly see its photons striking a solar panel. Yet the electricity generated can be easily measured and comes from a free source. With magnets as well, the source of free energy that replenishes at the atomic level is invisible to us. So perhaps a magnetic energy converter can be more easily thought of as a "cosmic windmill".

In explaining the source of power for new energy machines, Loncar starts with basics. Matter may appear to be solid and feel firm yet these are illusions for our limited senses of sight and touch. Instead, matter is made of rotating atoms, vibrating on a specific frequency, with a surrounding magnetic field. "Matter simply is an electromagnetic vibrating condition in space. Therefore energy already exists, it can only be transformed, but not be produced nor destroyed."[111]

While many researchers are using machines as a medium to transform energy, Loncar uses water as a medium. "The energy which we receive out of the water came from the ionization of the water. For a part of the ionization we have to pay, but the biggest part of ionization we receive for free by the vortex technology, manipulated by our electronic system."

A confusing array of names

The source of extra input into his device is the inexhaustible energy which pervades all matter, Loncar said. Some of the names people have given for that are:

Zero-point radiation
Radiant energy
Cosmic energy
Cold electricity
The sea of energy
Dirac Sea
Vacuum fluctuations
Higher dimensional energy
Zero point vibration
Residual energy
Quantum oscillations
Vacuum electromagnetic field
Virtual particle flux
Dark energy
Cosmic energy
Aether
Ether
Negative Electricity
Bioenergy
Orgone
Space energy field
Hyperspatial energy
Life energy
Creative vibration
Tachyon-energy
Prana
Chi
Scalar energy
Neutrinos
Quantum flux

No matter how we call it, it is a standing sound-wave which moves as a vortex," Loncar says. "The best condition for this vortex is the vacuum, or non-conductive, dielectric, non-linear materials."

A *standing wave* is one where the profile of the wave does not move through the medium it's in, but remains stationary. It results when a traveling wave is reflected back onto its own path. Loncar's view of invisible standing waves in space echoes the paradigm of a PhD frontier scientist across the Atlantic and other researchers. The *Space Resonance* Theory of Dr. Milo Wolff looks at empty space ordinarily regarded as "nothing" and says the properties of space are the basis of all physical laws, matter and the structure of the universe. He named a basic particle structure a space resonance "because a common term for a standing wave is a resonance and because these waves propagate in the medium of space." According to his concept, all the matter of the universe is made of space resonances. Other researchers want to know how to use that knowledge to build devices that work in harmony with such phenomena.[112]

Ervin Laszlo, PhD believes that standing waves are the mechanism by which the universal sea of energy records information about everything that's ever existed in it. He prefers the term "akashic field" for describing the information field of the cosmos.

> *A cave-dwelling early human must have been awestruck when he or she encountered a tree burning after it was struck by lightning. "Heat and light! How can I make that happen?" Our ancestors apparently started experimenting with anything that produces heat, such as rubbing sticks together. They didn't wait for physicists to explain friction or chemists to explain combustion as a chemical reaction. Later, the Neanderthals enjoyed their campfire quite happily without the knowledge that fire results from the exothermic oxidation of a hydrocarbon or carbohydrate.*
>
> *Similarly, why should the fact that scientists today disagree on the source and nature of background energy in space cause inventors to stop and wait until theorists catch up, publish papers and agree on a new science? That may take fifty years. We think that inventors should be allowed to bring practical working breakthrough energy technology to the public even while waiting for the definitive explanation of cosmic energy or whatever it will be called.*

Thomas Bearden is a theorist dedicated to helping energy researchers understand how to tap the potential energy in the zero point energy of the "vacuum" of space.

"In particle physics it is well known that the active vacuum is incredibly energetic," Bearden says. And in his view there are many ways to extract energy from that "vacuum." Calculations by establishment physicists [113] showed that a cubic centimeter of "empty" space, a volume about the size of the tip of a little finger, has so much raw energy in it that, if completely transformed into matter, the quantity of matter would be more than what is visible in the universe with the largest telescope! To most everyone, this sounds incredulous and paradoxical since it's hard to comprehend that what appears to be "nothing" (invisible space) is actually greater than a whole lot of what we consider "something" (visible matter). It simply points out that the vacuum of space is anything but empty—and also that the capability to tap even a tiny bit of that source would supply all the energy anyone might need.

Meanwhile in the electrodynamics model of thinking used in electrical engineering, the same scientific community assumes that the vacuum is inert. "The model assumes that all electromagnetic fields, potentials, and every joule of electromagnetic energy in the universe is produced by their associated source charges—right out of nothing at all, with no energy input to the charge at all, but with continuous energy flow from it." Bearden shakes his head ruefully at what he sees as inane contradictions. He writes articles prolifically, pointing out where electrical engineering has gone wrong and how it could instead be part of an energy abundance paradigm.

Resonance reigns

We've encountered the inward-spiraling vortex as a well-recognized pump for extracting energy from surrounding space, and there is another common theme in the emerging science of how to tap into abundant background energy. *Resonance* is basic to getting a large output of power from a small input, and is recognized as useful by everyone in the wide spectrum of viewpoints, even by the engineer who distances himself far from aether or zero point energy theories.

Everything, the largest and the smallest, has its own resonant frequency. Matter, including the bioelectric organism you call your body, is made up of subatomic particles that are in perpetual motion. The speed of the motion and the type of matter affects its "vibratory rate" or frequency.

Frequencies of other things interact with our own resonant frequencies. Everything in nature affects everything else. This is true for physical movements, for instance, but also for energetic movements both subtle and pronounced.

Perhaps you've heard the illustrative story that the flapping of a butterfly's wings in one part of the world creates effects that can multiply and alter the course of a giant storm on the other side of the world. That particular example is difficult to prove, as the effects may be imperceptibly small, but other physical effects involving a small input causing large output effects are proven in science every day.

Even events can resonate around the world. Psychically sensitive people living far away from New York felt the effects of major events such as the Sept. 11, 2001 attack on the twin towers. They felt the impact even though they had not tuned into a news channel and heard about it.

Music and sound are ancient forms of working with resonance. Like some of the unusual energy-converting inventions, music deals with oscillations, meaning vibrations of a part of an instrument at that part's natural frequency of vibration. If you hit a piano's 'A' key, a nearby violin's 'A' string vibrates.

An opera singer capable of shattering a wineglass with her voice would have the ability to project a musical note precisely; she would have perfect pitch. A glass made out of quartz crystal is more likely to be affected by vibration at the level of its internal structure than other materials. A loud sustained note of music would entrain the quartz' atoms whose natural rate of vibration interacts with that certain note, and when it is in resonance and the driving frequency suddenly shifts, the quartz material would break up.

Why is resonance a key to new energy science? It's because *atoms, sound, materials, magnetic fields, electrical circuits can resonate*. Around the world we meet inventors who make use of this small-input and large output principle to produce devices with exceptional yields.

You don't have to understand circuitry or complicated machinery to grasp the principle of resonance. Like a child on a swing, properly timed low-power nudges amplify into big results. You don't exert yourself and give one big push to your child's swing and walk away. Instead you stand there and tap on the swing at the precise time that momentum is enhanced. Momentum builds up gradually and the child swings higher and higher.

One inventor tells us that sudden expansion and contraction is key to resonance. "A pulse is an extremely, explosively quick expansion following after a contraction or compression." William Baumgartner, following in the footsteps of Viktor Schauberger, builds unusual devices categorized as "implosion technology". One of them is called the Wavy Plate. To get the pulsing action in the wavy-plate, its structure guides air or water through radial (circles one inside another) orifices which repeatedly compress the air or water and then let it expand and accelerate. Acceleration is repeatedly followed by deceleration. The temperature drops, similarly to what happens in a refrigerator where compression and expansion happen through only a single orifice.

Resonant geometrical shapes

Dan Davidson has expertise[114] in how *shapes* resonate with one another. Certain geometries entrain the natural preferred motions of the background energy of the universe. For example, in nature we often see the "hyperbolic spiral"—in sea shells, galaxies, fiddlehead greens, the arrangement of seeds in a pine cone and countless other instances.

The concept of what is called "sacred geometry"[115] implies that geometrical patterns exist everywhere in space and are related to the foundation of creation. Davidson says experimenters can observe physical-world results in their laboratory, tapping energy by using geometrical principles that exist in the natural order of the universe. "Certain shapes focus and transduce higher-order energies into electricity or magnetism." They can function as a natural generator.

There have long been claims that pyramids have unusual energetic effects, ranging from retarding the spoilage and enhancing the flavor of food and wine stored within them to improving the well-being of humans who interact with them. Could there be some real yet poorly understood energy phenomena that support such claims?

Davidson and his associates say they have proved that energetic effects come from the edges of pyramid shapes. When these shapes converge as angles, they create an aetheric vortex which translates aetheric force into magnetic and electrical forces. More research is needed to find the best angles if any, and to define the rules of how to convert power by using converging lines. Many energy-related physical laws are not understood right now. "What is now 'sacred geometry' will in the future become known laws of physics . . . . Maybe it should be called sacred mathematics."

The resonance project

Nassim Haramein of Hawaii is taking the emerging knowledge of both new and rediscovered science to the people, through both public appearances and widely circulated DVDs. With graphics and an engaging personality, he explains how the invisible aspect of the universe may be constructed out of very basic geometries in patterns that repeat at every size level from micro- to macroscopic.[116]

Decades ago, as a free-spirited young man who had escaped the confines of formal schooling, Haramein made his living as a ski instructor. Since he also relished mountain climbing, his early ponderings about the structure of spacetime took place out in nature. He realized that if he wanted to know if nature is being directed by its underlying structure, he should be able to see signs of it. Up in the mountains he peered closely at snow crystals and crystals in the rocks, looked at the way tree branches divide, thought about many aspects of the natural world and concluded that universal forces do continually repeat certain patterns. It seemed that the dividing of the underlying background energy always generates the same fundamental specific geometry.

Looking at the proliferation of spheres in nature and knowing that the sphere is the most unstable shape, he wanted to know what geometry holds the sphere together. The inner structure holding a sphere must be in perfect equilibrium, he realized, so that we think there's nothing there, just a vacuum.

His logic led to the tetrahedron, but there were many more puzzles to be solved before he reached the answer about a further geometric structure which meets all the requirements of his search—endlessly uniting the finite with the infinite, and uniting extremely small with extremely large objects.

His research resulted in a theory that unites universal forces and has implications in theoretical physics, cosmology, quantum mechanics, biology, chemistry, archeology and anthropology. For those who just want breakthroughs in energy science, it seems that Haramein's model of how the invisible universe works could help an experimenter to use specific geometries that would be more likely to resonate with the fabric of space.

Haramein proposes that instead of looking for fundamental particles, scientists should start looking for a fundamental principle of division, a fundamental *pattern* of creation. If we understand the pattern, the principle behind everything, Haramein says, we have the key to creation. Underlying geometry of space is that key.

A feedback loop between rhythmic contraction and expansion, the radiated side and the contracted side of universal structures sounds like a new field of physics, yet one partially predicted by visionaries such as the late Walter Russell (chapter 8).

The energy elephant

Perhaps scientists' disagreements about the qualities of the invisible source of energy is like the story of the blind men and the elephant. In that fable, each sightless man encountered one aspect of the elephant and concluded he knew what an elephant is.

The blind man who bumped into an elephant's foot and leg said "It's like a tree trunk." The man who reached out and felt the trunk said an elephant is like a hose that squirts water. The blind man whose hand found the elephant's tail said an elephant is ropelike. An elephant lying down presented a large flapping ear to the hands of another seeker. When the blind men congregated afterward, their definitions differed so wildly that they ended up angry and frustrated with each other.

Consider that story with "sea of energy" substituted for the elephant. Those blinded by the disease that Nobel laureate Brian Josephson calls pathological skepticism don't sense the sea of energy at all. You'll find them standing with their backs turned against the unknown, arms folded. Or they're ridiculing or from a safe distance hurling chunks of hardened dogma at explorers of the unknown sea.

Aether to zero-point

From the start of science, almost everyone believed in an aether or ether filling space and giving light waves a substance in which to wave. Then a famous experiment[117] done by Albert Michelson and Edward Morley in 1887[118] showed there was no "ether drift". (If the ether were unmoving, the Earth should feel a wind as planet races through the stationary ether. An ether would also change the way light travels.) From that time onward the scientific community has said, "There is no ether, and empty space is empty."

It was the most famous experiment-that-failed in history. Its failure came from the fact that Michelson and Morley assumed that the ether would be motionless in space. So they assumed that when the earth moves around the sun it must be *passing through* this motionless ether. As a result of Earth's motion, they figured, an ether wind must blow past the earth constantly.

However, their experiment only proved that an aether *wind* does not exist. Physicist James Clerk Maxwell mistakenly defined ether as a material substance, more subtle than visible bodies, existing in parts of space which seem to be empty. Frontier scientists in the new energy field view the ether as massless, not a material.

Those who took Michelson-Morley's lack of ether drift as proof there is no ether seem to assume that the ether would not be moving. Dr. Harold Aspden however has never assumed that the ether had to be something at absolute rest throughout the universe. "To me the aether is that unseen something that can store energy and affect light propagation but yet fills any free space in and around the protons and electrons and their derivative particles that we do see as the matter form," Aspden said in a 1997 tutorial on www.energy-science.org.uk.

Rolf Schaffranke (see Chapter 1, *Breakthrough Power*) told Jeane Manning that the theories of official science derailed when the absence of an ether wind was presented as the absence of an ether. Before 20th-century textbooks buried ether theory, science pioneer Sir Oliver Lodge said we have no way to grip the ether mechanically or move it in any ordinary way. "We can only get at it electrically. We are straining the ether when we charge a body with electricity; it tries to recover, it has the power of recoil . . . ."

The science giant relentlessly referenced by the etherless universe crowd, Albert Einstein, ate his own words a few years before his death. Schaffranke's book quotes a remarkable sentence from Einstein: "According to the general theory of relativity, space without ether is unthinkable . . ."[119] Einstein after his seventieth birthday wrote to a friend, "You imagine that I look back on my life's work with calm satisfaction. But from nearby it looks quite different. There is not a single concept of which I am convinced that it will stand firm, and I feel uncertain whether I am in general on the right track."[120]

Miller surpasses Michelson-Morley

Unknown to most science students today, however, a colleague of Michelson, Dayton Miller, who worked with Michelson and was president of the American Physical Society went to the trouble of doing expensive ether drift experiments using an interferometer that was the most accurate such measuring instrument in the world at that time. He did the work for over 20 years, making thousands of measurements in all seasons. In contrast, Michelson and Morley only did four six-hour runs in the same week at one location. Miller did measurements at sea level and also on the top of a mountain.

At sea level Miller discovered results similar to Michelson/Morley's, but the ether drift was more detectable up on Mount Wilson. Miller's results varied as the earth rotated. His experiments supported a model of the ether being dragged along with Earth's motion.

James De Meo, PhD looked into the Dayton Miller story and found scientific banishment and politics directed against Miller to minimize the influence of his experiments.[121]

Fuel of the future

Are concepts of ether and zero-point energy interchangeable? Some researchers say they are. The late Eugene Mallove, PhD[122] thought so until he had collaborated with Paulo Correa, PhD and Alexandra Correa who performed successful experiments that supported the lifeforce concepts of the late Wilhelm Reich.[123] Mallove, whose background had been in mainstream science, had witnessed demonstrations of over-unity (more output than measurable input) from the Correas' devices such as the Pulsed Abnormal Glow Discharge. He even experienced interactions between the invention and living-energy fields such as those from his hands. His conclusion was that zero point energy is a limited model because it does not explain anomalies such as the excess heat from "cold fusion".

Thomas Valone PhD is a professional engineer who stays within accepted terminology. He researched examples of promising experiments related to zero point energy and found an abundance, such as Franklin B. Mead Jr., a scientist with the US Air Force, who received a landmark patent for a way of "converting high-frequency zero point electromagnetic radiation energy to electrical energy."[124] Valone also has a colleague who left an impressive science career to develop the field of quantum vacuum engineering. Fabrizio Pinto PhD worked at the Jet Propulsion Laboratory on the space program, was a senior astrophysics lecturer at the University of Southern California and left those circles to start InterStellar Technologies Corporation.[125] "In the next few years we will see changes in science and technology that will make Star Trek scripts look short-sighted and obsolete," Pinto says. In the last few years his company has applied for patents in the area of revolutionary means of energy production, citing zero point energy. "This means measurable progress in the quest for fuel-free, by-product free energy, including much more efficient approaches to traditional technologies."

At the same time, prominent science writers call such efforts "pseudo-science that could leach funds from legitimate research. The conventional view is that the energy in the vacuum is miniscule."

Aside from the diversity of viewpoints about this mysterious energy, is *zero-point energy* the best moniker for the background energy of the universe? Quantum physicists call it zero point energy because even at the theoretical lowest temperature ("absolute zero", minus 273 degrees Celsius), at which point the absence of heat would be expected to result in no molecular motion, motion (thus energy) still is present. But doesn't the phrase "zero-point" sound negative and inadequate to describe the unlimited hidden power plant of the universe? "Zero" carries the baggage of "insignificant, zilch, nothing, nada".

One mainstream journalist wrote that zero point energy is *produced by* the miniscule movements of molecules at rest.[126] If that were a true definition, then it would indeed be a miniscule source of energy. To the contrary, zero point energy *produces* such jittering; it is a cause and not a mere effect.

Today so much misinformation is splattered onto the internet in the name of zero point energy that some of its advocates turn away in embarrassment. More than one leading investigator has taken to calling it the virtual photon field or virtual photon flux instead of zero-point energy. We find physicist Bernard Haisch's phrase "a background sea of light" more appealing than zero-point field, but he too adds the quantum physics descriptor for it—electromagnetic zero-point field of the quantum vacuum.

Even that view of the background energy's qualities is controversial. Some scientists in the new energy field say that trying to fit new science knowledge into the standard twentieth century accepted body of science just won't work. They question whether the energy embedded in the background is really electromagnetic, or if it's off the scale of the electromagnetic spectrum as we know it, or something quite different such as a living but non-material field with intelligence.

The scientists we meet in the next chapter avoid such speculation and work with accepted scientific principles and terminology. Nevertheless they are considered heretical because of the possibility that they could bring humankind cheap power for heating homes and industries.

12

The Water is Hot

Emerging technologies point to a much brighter future for mankind. They do not require resources controlled by any small group of countries. They are portable and democratic.

—*EUGENE MALLOVE, ScD in report on request from the White House.*[127]

Sonofusion has the potential for the complete replacement of CO_2-producing fuels. This would have world-changing consequences.

—*ROGER STRINGHAM, PhD sonofusion pioneer*

These (new energy breakthroughs) take a huge amount of effort, and if you have somebody throwing rocks at you all the time, it's hard to muster the effort that's needed for advances.

JOHN DASH, PhD cold fusion researcher

Industrious and clever entrepreneurs rise from obscurity and a humble lifestyle to respectability, fame, and financial fortune, according to the rags-to-riches folklore. However, the reality is that if they made heroic advances in science or engineering that hold promise for replacing polluting hydrocarbon fuels, their life stories have often been far less magical. Pioneers of breakthrough energy technologies have frequently encountered obstacles beyond those of a mere technical nature. Man-made hurdles have been placed in their paths, challenging the endurance, fortitude, and resources of even the most devoted and credible scientists. Nowhere is this more evident than in the cases of those who have discovered how to coax fire from water.

Bubbles in paradise

In Hawaii as this is written, gray-bearded physicist Roger Stringham and his wife Julie are packing up his research materials and the scientific instruments in his garage laboratory. The tall lean couple may look like those care-free retirees who leisurely spend the Kauai afternoons on the golf course or canoeing around the island where they live, but their lives are not so worry free. The savings and investments that have sustained them to this point are now dwindling, and today they are selling their home and seeking a more modest dwelling. This economic measure is a consequence of decisions they made years ago to support Roger's scientific passion. Stepping away from well-paid employment as a staff scientist at a prestigious research institute, he has devoted himself to full-time work on his energy breakthrough.

Roger Stringham's colleagues in an international but small scientific community recognize his energy innovation as potentially world-changing, but few outside that circle have heard of "sonofusion".[129] The compelling contents of the scholarly papers and presentations Stringham has put forth at a large number of international science conference have not yet made their way to the mainstream.

What is so fascinating about Stringham's discovery? Basically, he has harnessed the amazing power of microscopic bubbles to release surprising amounts of energy when they implode in specially treated water. Called "transient cavitation bubbles", the tiny yet powerful shock waves produced by the collapsing bubbles are well-known for common nuisance effects, but less known as a potentially potent new source of clean energy if properly harnessed.

On a larger scale, you may be familiar with cavitation as the "water hammer" that loudly vibrates household plumbing. If the pressure on a fast-moving liquid drops suddenly, tiny bubbles ("cavities") of vapor may form spontaneously in the liquid. When the bubbles collapse they unleash tiny shock waves with enough power to rattle water pipes or cause destructive pitting in the metal of boat propellers and pump impellers. Roger's technical background positioned him for finding a beneficial use for cavitation. Previously as an industrial consultant he had dealt with the problems that cavitation causes for the marine and other industries. In his work now he intentionally induces cavitation to capture the energetic effects of the collapsing bubbles.

Ultrasound makes waves

In his experiments Stringham uses ultrasound—frequencies too high for the human ear to hear—to pulse rapid pressure waves through small quantities of liquid deuteriumoxide, the nonradioactive "heavy water" from an isotope of hydrogen. That shaking of the liquid causes microscopic voids to form and then violently collapse. Roger found that when the bubbles collapse they shoot a tiny high-velocity stream of hot ionized water vapor a short distance through the liquid, what he terms a "plasma jet". When the plasma jets impact the surface of a special metal foil, deuterium nuclei are forced into the metal, resulting in a reaction that releases a tremendous amount of heat energy that quickly raises the temperature of the surrounding liquid. Electron microscope scans of the metal foils show microscopic "volcanoes" where molten metal erupted out of the foil.

Stringham's experiments done in a darkened room are illuminated by blue or ultraviolet light. Blue light flashes associated with liquid excited by sound waves has been studied throughout the world under the name sonoluminescence. Stringham's breakthrough with his sonofusion was the fact that he demonstrated a non-radioactive nuclear reaction process for reliably generating significant excess heat economically. The byproduct is helium, the same gas pumped into balloons for children's birthday parties.

He also has dramatically scaled down the size of his reactor while keeping the same heat output. Increasing the frequency of the ultrasound generator to 1.6 million cycles per second and working with the resonance factor contributed to that accomplishment.[130] The reactor is the size of the timepiece of a man's wristwatch and weighs only twenty grams (0.7

ounces). Its heat output of forty watts would only power a small light bulb, but the small devices could be ganged together to form a battery-like device of any size with a high density of energy output. Each unit is bathed in water to remove its excess heat quickly so the device won't over-heat.

If Stringham and his business partner could get financial backing to hire a team of specialists and buy the materials and equipment needed for a more rapid succession of experiments and engineering refinements, progress toward a product manufactured for the marketplace could be dramatically speeded up.[131]

Stringham's repeated experiments show that sonofusion is ready for that commercial level; it could be heating large buildings where people shop and work. Space heating uses up nearly a third of the electrical grid power in the United States. Farther in the future, he believes, sonofusion could replace carbon fuels. "This would have world changing economic, political, and environmental consequences."

Heavy water: abundant, cheap, nonradioactive

The "fuel" Stringham uses is found in ordinary water at about one part in 6,000. He says it has potential for a million years of the world's energy supply. "The amount of energy that can be extracted from D_2O is one million times that of gasoline. Very high grade deuterium costs fifty cents a gram and this one gram is more than enough to power a car for its lifetime . . . The car will wear out before running out of its fifty cents of fuel."

The world won't run out of water from using deuterium because this form of hydrogen makes up only .015 per cent of the hydrogen found in ordinary water. Looking at sonofusion on a smaller scale, a gallon of ordinary water contains enough heavy hydrogen to produce the energy-equivalent of 300 gallons of gasoline.

Here's what you need to know about hydrogen: It is the lightest and most abundant element in the universe. Hydrogen is a colorless, odorless gaseous chemical element that comes in three varieties or "isotopes"—ordinary hydrogen (protium), deuterium and tritium. [Tritium is made artificially, but the other two isotopes of hydrogen occur in nature.] Chemically they are identical, but the hydrogen triplets are distinguished by atomic weight. Ordinary hydrogen is the lightest and tritium the heaviest isotope of the three. "Isotope" is defined as one of two or more atoms of the same element that have different numbers of neutrons. Ordinary hydrogen has one proton and no neutrons in its nucleus, deuterium has one proton and one neutron, and tritium has one proton and two neutrons. Ordinary hydrogen is by far the most abundant type, comprising 99.94% of the total hydrogen found in water. But deuterium fuses much more readily, and therefore is the preferred fuel in a fusion reactor.

Bubbling controversy

Although the formation of cavitation bubbles is inherent in both processes, sonofusion is different and apparently more commercially viable than the "bubble-fusion" of Rusi Taleyarkan that uses a deuterium-containing organic solvent to generate heat from the fusion of deuterium nuclei. Stringham's sonofusion produces many orders of magnitude more heat than Taleyarkhan's process in the same size apparatus.[132] But although its output is much less than from sonofusion, it has predominantly been the politics of science that have hampered bubble-fusion's own progress. Described as hot fusion on a small scale, bubble-fusion ignited a storm of controversy in science journals a few years ago. After years of careful laboratory work, Oak Ridge National Laboratory scientist Rusi Taleyarkhan suddenly became the target of allegations which turned out to be unfounded and unethically motivated. His name was cleared repeatedly, then again tarred with insinuations in popular science publications.[133]

Meanwhile, "hot fusion" projects have received billions of dollars of public money over the years, but its well-worn promise to supply the world's energy needs has for decades been a moving target—usually fifty years down the road.

Before we look further into the politics of the field, let's first answer the question "What is fusion?"

Fission vs. Fusion

Fusion is the opposite of fission, although both processes start with atoms. Atoms are the tiny building blocks that make up all matter. An atom consists of a nucleus—which is made up of protons and neutrons—and electrons, which form a cloud around the nucleus. The atoms of each individual chemical element contain a specific number of protons, neutrons, and electrons.

Fission is the splitting of the nucleus of a heavy element, such as by bombarding it with neutrons. This creates lighter elements and releases a great amount of energy—heat, light and various forms of radiation. Radioactive materials with unstable atomic nuclei, such as certain isotopes of uranium or plutonium, are needed for fission. Atomic bombs such as those dropped on Hiroshima and Nagasaki, and today's nuclear power plants, generate their energy using nuclear fission.

Fusion is the joining together of atomic nuclei of lighter elements to form a heavier element. Hot fusion, which is said by mainstream scientists to be the process that energizes our sun, uses for fuel the deuterium form of the lightest element, hydrogen. Textbooks teach that temperatures reaching millions of degrees Fahrenheit are needed before the positively charged hydrogen nuclei can overcome their natural repulsion from each other, since like charges repel (think of what happens if you attempt to bring the north poles of two magnets together). If the deuterium nuclei do come close enough together, they form a slightly heavier element—helium. And in the process, an enormous amount of energy is released in the form of heat and light. Some harmful forms of radiation may be released as well. Hot fusion is the source of energy for the devastating power of modern thermonuclear weapons, so-called "hydrogen bombs".

Hot and cold

If an economical and safe way to generate energy from fusion can be developed, it would indeed be an historic breakthrough. However, current experimental hot fusion technology is an enormously expensive process which needs tremendous amounts of heat and pressure, and huge reactors. Despite decades of research and billions of tax dollars spent, experimental results have been disappointing time after time. Any plan to build a practical hot fusion-based nuclear power plant is chronically decades away.

In contrast, Stringham's sonofusion reactor does *not* produce harmful radioactive byproducts.[134]

Cold fusion refers to the joining of atomic nuclei at normal room temperatures.

Each pair of heavy hydrogen nuclei fuses to form one helium nucleus and the process releases heat.[135] Instead of hot fusion in superheated gas, cold fusion is usually based on reactions helped along by a metal catalyst such as palladium, which has relatively large spaces in its microscopic lattice structure. It is theorized that deuterium packs into the spaces within the palladium in the same way that water moves into the open, absorbent surface of a towel.

Nuclear physicists who are experts in hot fusion expect any fusion reaction to be accompanied by a spray of radioactivity. For years cold fusion proponents could not prove that the heat generation in their experiments is the product of actual nuclear reactions, due to the absence of the expected radiation. Thus, most conventional scientists rejected the whole idea of cold fusion. Stringham's discovery may have been caught in the fallout from the controversy because his method of injecting deuterium plasma into a metal lattice is said to trigger cold fusion.

Name change

Even in the cold fusion field some scientists who have witnessed successful experiments are questioning whether fusion of deuterium nuclei is what is causing the excess heat. At the time of this writing the question remains open.

Rather than the phrase cold fusion, the terms *low energy nuclear reactions* (LENR) and *chemically-assisted nuclear reactions* (CANR) are used by some scientists who do the experiments. They see it as part of the field of condensed matter nuclear science. Increasingly, science writers use the LENR name.

Name-calling

The branch of scientific research mislabeled "cold fusion" is like a teenager who has been called "impossible" for nineteen years. Angry neighbors in the science community muttered "fraud" and publicly directed insults at the troublemaker. "Voodoo science!" Most people in the community shunned the troublemaker after hearing the critics' name-calling repeatedly through the years.

Now fast approaching maturity, the 19-year-old is delighting in proving to those critics that they're wrong. A recent announcement in Japan and discoveries at a United States Navy laboratory appear to vindicate the validity of the rebel technology.

Born into fame, disgraced, redeemed

On March 23, 1989, University of Utah chemist Stanley Pons and British electrochemist Martin Fleischmann from the University of Southampton in the United Kingdom shocked fellow scientists when they announced at a Utah press conference that they had fused atomic nuclei in a jar of water, in a tabletop device they had made themselves. They had done this at room temperature by immersing a palladium rod and charging it with a small amount of electricity. As we have seen, fusion is usually expected to require the high temperatures of a sun. Another surprise was that they found far more energy was released than could be explained by typical *chemical* reactions, which meant a more powerful *nuclear* reaction could be the explanation. Yet there was no evidence of harmful radiation.

Their experiment involved electrolysis of heavy water, although later experiments used ordinary water. The heavy hydrogen packed into palladium's lattice structure so closely that it mysteriously overcame the usual repulsion between the positively charged protons.[136]

Around the world scientists rushed to replicate the experiment, but trouble soon sprang up from roots anchored in the politics of mainstream science.

Pons and Fleischmann had bypassed the usual route for new scientific discovery, which is to send their findings to an established science journal for publishing so that other scientists can critique their methods and try to replicate the results. This standard procedure is called the "peer review" process. (Going straight to the press was not their preference. However, their employer had pushed them into making the announcement at a press conference in order to upstage a competing scientist at another university, but critics perhaps did not know that.)

Some physicists also viewed the two chemists as outsiders intruding on nuclear physicists' territory—fusion. Physicists elsewhere did not necessarily know electrochemistry techniques and also did not allow enough time for the heat-generating reaction to proceed. Fleischmann and Pons had not told them that preparing the materials is very difficult. When less rigorous researchers failed to duplicate Pons and Fleischmann's results, they dismissed the experiment as bogus . . . Also, since Pons and Fleischmann did not yet fully understand how they had accomplished "excess heat" from the experiment, the recipe they gave was incomplete.

As we've seen, hot fusion scientists claim that fusion of nuclei always releases harmful, if not deadly, radioactive byproducts. The fact that Pons and Fleischmann were still alive was used as an argument against their claim of nuclear reactions.

The hot fusion camp was particularly defensive of their territory because the United States Congress at the time was reluctant to renew the annual budget for hot fusion research. By then the government had spent $15 billion without hot fusion having achieved even an energy breakeven point. Heavy vested interests in the research grant gravy train were at stake.

In 1989 the United States Department of Energy (DoE) hurriedly brought together a panel (a bevy of hot fusion scientists) to decide the fate of cold fusion, and later that year decided not fund the new research area. The next year Pons and Fleischmann gave more complete details but by then the media was not listening to them; cold fusion had been declared dead. Their black-balled treatment by the mainstream scientific community drove the two scientists into the equivalent of exile, and they resumed their research at a laboratory in France funded by a Japanese corporation. Eventually Stanley Pons left the cold fusion field entirely. We speculate that he and his family decided they had suffered enough derision.

Contrary to popular opinion, scientists in thirty respectable laboratories around the world continued, with their own time and money, to refine the processes, in some cases replicate the experiment and make further advances. Generally it was "back-bench" work, using the laboratories' equipment after hours.

John Dash and student with cold fusion cell.

One of the unsung success stories

At universities farther away from the politics of hot fusion, researchers with a degree of independence were determined to advance the frontiers of science. They also had other motives. If cold fusion could be made reliable and totally safe, the excess-heat output could be used for homes, greenhouses, steam turbines and in general improve people's lives. It would also shift the balance of political power in the world, because the fuel—water—is available to everyone.

At Portland State University in Oregon, the head of the physics department heard the news and asked one of his professors to study the Pons-Fleischmann paper. That professor is a quiet, careful researcher, Dr. John Dash, a metallurgist experienced in electroplating. Dash tried the Pons/Fleishmann experiment and was surprised. The metal distorted as soon as he started. The distortion was not caused by heat, but by internal stress produced by absorbing deuterium. Dash had never seen anything like it. "So I was hooked," he told us.

Over the following years Dash brought young people into the cold fusion field by involving his students in it. He was described, by an engineer who visited him, as thoughtful but stern and insistent that students build consistent evidence in their experiments.[137] Another science journalist describes him as so professorial you would think he was sent by central casting.[138] Early in the saga one of his students said, "Dr. Dash . . .despite being underfunded and understaffed . . .has managed to put together working cold fusion cells. While these cells do not always produce excess energy indicative of cold fusion, they do often enough to provide promise."

With the help of the Drextel Foundation and a couple of grants[139], Dash was able to have an ongoing program of teaching cold fusion work to summer courses of high school and college students, with the help of two graduate students. Eventually Dash changed his focus from the Pons/Fleischmann process to using titanium and platinum, and heavy water doped with an acid. He could continue seeking an answer to the mystery of exactly why excess power comes from cold fusion cells only under certain conditions. How could the experimenters make it happen more reliably? And there was a further mystery to solve—the apparent alchemy taking place in the cells. A metallic element changing into another element in a laboratory was too much for many of Dash's colleagues to believe.[140] Dash and others are in the middle of yet another scientific upheaval—they are getting transmutations of elements in their experiments—another clear sign of nuclear reactions.

Mallove informs President

The late Eugene Mallove PhD remarked on the fact that the experimenters cannot yet make a cold-fusion machine that you can buy in a chain store, but do have peer-reviewed science papers about positive results in their field. That fact "is no more condemning of cold fusion than it is that in 1947 we had the transistor but we didn't have the transistor radio or computers yet".

Mallove was an aerospace engineer with advanced degrees from Massachusetts Institute of Technology (MIT) and Harvard University and author of five science books. In 1989 he could have remained as MIT's public relations officer and chief science writer and had a financially comfortable career, but his conscience would not allow him to ignore MIT's unscientific and politically-charged treatment of the fledgling science that was misnamed "cold fusion". He left because he believed that a scientist there had changed the graphed results of an experiment in order to hide the fact that MIT had indeed seen excess heat coming from a cold fusion cell—a clear breach of science ethics.

After leaving MIT, Mallove turned to writing about cold fusion and co-founded a magazine, *Infinite Energy*.[141] During the administration of United States President Bill Clinton, Mallove was asked by White House staff to provide an essay on cold fusion as part of a future-oriented book for Clinton. In his summary Mallove told what it would take to get low-cost new energy products widely distributed in the marketplace. It will require removal of internal opposition from vested interests in the U.S. government and industries, including the blocking of 'cold fusion' patent applications by the government's patent office, he told the president.

LENR successes abundant

Eugene Mallove had the gratification of seeing successful demonstrations run several days[142] at MIT during the tenth International Conference on Cold Fusion. It particularly warmed him because MIT is the alma mater of both himself and Dr. Mitchell Swartz, who brought a working cold fusion demonstration. Mallove said, "That was within a few thousand feet of the Tokamak hot fusion laboratory which has consumed something like approximately a half billion dollars of federal funds over the last 15 years. The hot fusion graduate students have learned this or that but the program is going absolutely nowhere. And here . . . was a working cold fusion reactor . . . of significant and very accurately measured performance."

Reproducible production of energy has also been shown at experiments replicated at laboratories ranging from Israel,[143] to an Italian government laboratory in Rome, Italy, and SRI International in Menlo Park, California.[144] In May of 2008 Osaka National University professor emeritus Yoshiaki Arata demonstrated his successful cold fusion reactor for Japanese media and international guests. He is one Japan's leading scientists, with honors including a hall named after him at the university and an award from the Japanese Emperor.

The website www.lenr-canr.org has a library of more than 500 original scientific papers, linked to a bibliography of over 3,000 journal papers, news articles and books about the field formerly called cold fusion. Some of its practitioners are being invited to present their findings to science conferences such as the American Chemical Society and The American Physical Society. Roger Stringham is one of those invited to present his findings at prestigious meetings this year.

"There is a substantial amount of work to be done to make sonofusion a replacement for a hydrocarbon-based economy," Stringham reminds fellow scientists.[145] That needed work could pave the way for the other safe-fusion approaches. A task force of scientists is needed to find new and better thermoelectric devices for directly converting heat to electricity. (Potential breakthroughs brought to the authors' attention by other inventors hold promise for very efficiently converting excess heat into electricity. Ken Rauen for instance proposes a radically different approach to using heat. Rauen's "superclassical heat engine" concept is designed to tap the heat of the environment as its heat source.[146] Rauen worked on replicating Stringham's sonofusion and is convinced that the sonofusion process does produce excess heat.)

Breakthrough at Navy lab

Strong and repeatable evidence vindicating Fleishmann and Pons' claim came from the U.S. Space and Naval Warfare Systems Center (SPAWAR) laboratory in San Diego, California. A research team[147] had developed a procedure which deposited palladium and deuterium together on a thin metal film.[148] In 1995 they first found indications of nuclear activity. A decade later they had further evidence and also discovered that reactions could be speeded up and new elements produced by placing their electrolytic cell in an electrostatic field.

A simple piece of plastic placed next to the cathode caused further triumphant moments at the SPAWAR naval laboratory. That plastic piece is a detector[149] of charged particles and is used regularly by experts in hot fusion. The team is employing it to show evidence of low-energy nuclear reactions in their palladium lattice and charged particles emitted in amounts far greater than the background level. The neutron tracks registered by the detector are so numerous that these scientists say they provide undisputable evidence of the tracks' nuclear origin.

Tritium not included

There are no radioactive byproducts at all from Stringham's or many of the other varieties of low energy nuclear reactions (LENR) experiments. Some others, however, see traces of tritium when they analyze what is left after the experiment.

Tritium is a radioactive form of hydrogen and is used by medical researchers. Tritium's radiation is so weak it cannot penetrate human skin but it is readily detected. It has a half-life of 12 years and can be shielded by a piece of paper, but you don't want it in your drinking water because bodily tissues would then be adjacent to the weak radiation.

Some of the "cold fusion" experimenters seem pleased if they see traces of tritium when the aftermath of their work is analyzed. They may feel vindicated in the eyes of their scientific peers, because tritium is one of the expected byproducts from a nuclear reaction.

Other scientists in the cold fusion/LENR field celebrate the fact that their own methods of getting excess heat are tritium-free. Dr. Mitchell Swartz points out that when plasma is injected into a metal lattice it's a different game than hot fusion. "A mechanism exists that causes fusion at a much greater rate and without neutron radiation and tritium production."[150]

The experiments are difficult to reproduce as long as scientists don't agree on what is really going on in the unusual processes. Around the world scientists continue to witness evidence of something new and unexplainable. The main byproduct is helium, not a greenhouse gas. The notable excess energy output cannot be explained by typical chemical reactions, and "cold fusion" does not look like any previously known nuclear process.

Not sure it's fusion

"The matter of fusion remains unproven, but after two decades clear evidence for some new kind of nuclear process has emerged," writes New Energy Times editor Steven Krivit.[151]

Ten years after the original Utah announcement, an energy historian compared the then-inconsistent results from cold fusion experiments to learning to build a fire. "Heat flares up and gutters out, like flames from green wet firewood."[152] However, he expected that when researchers learn the keys to controlling the reaction, they can scale up the most benign and the best approach to cold fusion. "We will not scale up the uncontrolled on-again off-again heat, or tritium production."

In other words, he expects that when scientists understand what is happening in "cold fusion", processes that have as a byproduct even a slight amount of radioactive material will be dropped. "Once we learn how to build this new kind of fire, we will make only clean, hot reactions."

13

Better Than Fire

Researchers at the National Aeronautics and Space Administration (NASA), scientists for the U.S. Navy and a group of investors who have contributed over $25 million to this private company believe in the work of Harvard Medical School graduate Dr. Randell Mills.[153]

Hydrino energy is a radically new form of energy, so radical that it has met intense resistance.

—THOMAS E. STOLPER, *author of Genius Inventor*[154]

Almost always the men who achieve these fundamental inventions of a new paradigm have been either very young or very new to the field whose paradigm they change.

—THOMAS S. KUHN, *author of The Structure of Scientific Revolutions*[155]

Imagine powering all of humankind's industries, homes and vehicles with a source of clean fuel that is so familiar and so abundant that it falls from the sky–water. Dr. Randell Mills once said that using this power is possible to do,[156] and as this is being written his company's related announcement is stirring up a frenzy in the realms of science and among those who watch energy developments closely.

Mills is CEO of BlackLight Power Inc., which announced[157] it has successfully tested a new energy source. The unique BlackLight process of extracting high levels of thermal energy from the hydrogen in water is neither nuclear fission nor fusion nor combustion. The technology generates no radioactivity and no other pollution. Instead, years of mapping the behavior of electrons in atoms and molecules led to the company's invention that incorporates hydrogen and a metallic catalyst into solid fuel. When heated, both hydrogen and the catalyst are released as a gaseous mixture that begins an unusual energy-releasing chemical reaction.

The prototype of Mills' system generated fifty kilowatts of thermal power on demand, and the projected cost per kilowatt for equipment and installation for a power station is less than half the cost of the nearest competitor, an industrial gas turbine power plant. By using the heat from the process to generate electricity, BlackLight Power estimates it can produce commercial electrical power at the unprecedented cost of only 1 cent per kilowatt-hour, which is much lower than even the most efficient coal-fired power plant.[158]

The BlackLight Process has no fuel costs, because the hydrogen is generated from ordinary water, and the spent catalyst recycles into more solid fuel. Therefore the system holds promise as a cheap supply of abundant heat to create the steam that turns turbines and generates electricity, and if the process is scaled up and mass-produced it could replace all the polluting power sources.

Further, the company said the BlackLight Process is ready to begin the leap into the marketplace. The company was in talks with engineering and manufacturing firms and projected a pilot plant could be built in 18 to 24 months.

Unique theory, unique new compounds

To say that Randell Mills aims high is an understatement. His vision encompasses the cosmos. This scientist/inventor/medical doctor believes he corrected the accepted theory that underlies today's physics texts. And while writing his "unified theory of everything",[159] he discovered how to make unique new materials.

Mills was still in his twenties when he began the calculations for his theory. As a scientist does, he tested it repeatedly by doing experiments. At one point he pored through books to find which *catalyst* (a substance that initiates and accelerates a chemical reaction without being changed itself) would fit his numbers. Experimenting led to making new chemical compounds.

Eventually he developed a process (at the time named HydroCatalysis) which used a common element[160] as a catalyst and freed impressive amounts of heat energy from water.

Leap past the steam era?

Mills said in the year 2000 that his team could probably skip a couple of generations—pass by the expensive steam turbine technology and progress to a direct *plasma-to-electricity* process. Plasma is a mass of hot ionized electrically-conductive gas, and Mills' technique apparently generates such a substance out of his solid fuel. He said Russian scientists developed plasma-to-electricity technology in the 1960s. And while BlackLight Power experimented with that idea, ultimately they decided to ease into the market place by coupling their breakthrough with standard power generation principles—adapting their innovation to the heat-to-steam process used in existing power plants. BlackLight would then also be in the "retrofit" business.

Their technology could conceivably replace the huge baseload of fossil and nuclear-fueled electrical generation. They particularly look to the smaller efficient heat engine called the "Stirling engine" for using their heat output to generate electricity. The Stirling engine and turbine hold promise of providing mechanical and electrical power, heating and cooling. However, the "dynamic plasma-to-electric converter device" is still on his list of applications for the discoveries.[161]

Replacing gasoline

"If you make cheap heat, you can make cheap electricity, and if you can make cheap electricity you can make cheap hydrogen," said the press release. In other words, generating electricity is not the only way to benefit from Mills' discovery. "The hydrogen-burning car has been possible for decades, but there has never been a way to produce cheap hydrogen—until today,"[162] Mills said. A part of the thermal power generated by the BlackLight Process could be used for making enough electricity to electrolyze more hydrogen out of water.

Mills figured his company could scale up the prototypes within two years to the point of having a supply of hydrogen that would replace the fuel pumped in a day at a gasoline station.[163] We hope his opponents step aside and let that happen.

H_2 Gas Station.

Lessons from farming

In his youth, Randy Mills developed a work ethic and a down-to-earth outlook as well as a curiosity that ranged out to the stars, according to a biographer.[164] Living on his family's dairy and grain farm in Pennsylvania and harvesting hay and corn from his own leased acreage while in high school taught him to be resourceful and undaunted by setbacks. Farm life had aspects in common with science, he noted later[165]—mechanics, chemistry, biology are an everyday part. Mills observed early in his life that nature is rich and complex but also understandable, predictable and unified.[166]

The six foot five inches tall, energetic young farmer was making money by growing grains and had no plans to go to college, despite having been considered the class genius in high school, but his outlook on life abruptly changed when he fell into a glass door. The accident, and the six hours of surgery needed to repair his hand and arm, shook up his thinking. While hospitalized from loss of blood, he realized that since eventually he would die, he first wanted to learn what he was doing on Earth—and how everything works.[167]

Mills had enough money in profits from farming to pay tuition at a local college in Lancaster, Pennsylvania, and he graduated at the head of his class. Taking as a role model the surgeon who had saved his life, he went on to Harvard Medical School. His plan, however, was to develop new medical technologies rather than practice medicine.

Harvard and MIT

Attending Harvard Medical School unleashed a torrent of directed creativity. For the inquisitive young man it was like being in a candy shop. "At Harvard you could be merely a student and pick up the phone and talk to anyone; they'd talk to you."

In addition to accessing intellectual resources in the Boston area, he could borrow measuring equipment for experiments. And he could regularly drop in at Harvard's Office for Technology Assessment with his inventive ideas.

One day a man who worked in that office said to Mills, "I've got to introduce you to someone." Mills was ushered into the medical school office of Dr. Carl W. Walter with the introduction, "This guy's coming in all the time with ideas . . . . Dr. Walter, you're a great inventor; I thought you two would hit it off."

They did, Mills said. "I wouldn't be here today if it wasn't for his influence."[169] The professor was a mentor even in his magnanimous attitude toward society. Walter in his own youth had worked his way through school, had a mentor and decided to give back to life. He was a surgery professor who could also teach his protégé about business because he had founded a dozen companies. At that time Mills was inventing for the medical field, and Walter put together Mills' first contract with interested parties in Japan, involving pharmaceuticals and new cancer-therapy technology.

Mills' mentor bolstered his confidence. "When you're in Harvard Med you feel a little overwhelmed." Mills remembers his feeling of shock one day when his mentor was speaking to very accomplished colleagues and predicting that Randy Mills would go down in history as one of the most successful scientists of the last few hundred years.

During his last year at Harvard, the ambitious and curiosity-driven Mills also studied electrical engineering at Massachusetts Institute of Technology (MIT).

Kitchen collaboration

Mills' energy breakthrough resulted from work he had been doing at MIT.[170]

He had received a paper containing the mathematics for a type of laser[171] from a professor[172], and reasoned that he could similarly apply mathematics to a certain problem of atomic physics.[173] Mills did, and believes that his solution eventually allowed him to solve physics problems from quarks to cosmos as well as to make the new materials. The theory he began to develop at MIT is said to describe an electrical engineer's universe—ruled by the classic Maxwell's equations for electromagnetics.[174]

After graduating from Harvard, he returned to Pennsylvania to work on his inventions,[175] partly because an academic career could well require conformity of thinking. Medical and energy devices were of equal interest to him. Eventually he was running his electrolytic cells in the kitchen of his apartment, collaborating long-distance with a team of scientists from General Electric who had developed a method for reliably measuring heat in such a cell. Mills faxed his numbers, and they in turn commented on his energy-balance equations.[176]

"It looked good to them, and I ultimately published an article and had a press conference in the county courthouse lobby in Lancaster."

Local company

Among those interested in the young man's work was a local company, Thermacore. A team from Thermacore tried Mills' experiment. At first it didn't go well, but with persistence it began working very well. Mills recalled, "We were getting ten times the power out, relative to the input power. It ran for about fifteen months."

A scientist[177] Robert Shaubach from Thermacore announced the work at an engineering conference in San Diego in 1992.[178] An undertone of excitement rippled through the room when he said that Dr. Randell Mills had achieved excess heat from an experiment using ordinary water as a source of hydrogen.

Controversy

However, his theory involving an unusually small type of hydrogen atom he calls a "hydrino", in which the single electron orbits closer to the atomic nucleus than previously thought possible, is unacceptable to scientists who believe that the hydrogen atom cannot achieve such a low "ground state". *Ground state* means the lowest possible energy level of the electron or electrons that orbit an atom's nucleus, and correlates to the minimum average radius of the electron orbit. Mills' claim that he can induce hydrogen atoms to a lower energy state is impossible if conventional chemistry and physics textbooks are correct.

In his theory about the creation of that smaller atom, the transition of the electron from its typical ground state to the lower hydrino ground state releases energy, primarily in the form of heat—as in the BlackLight Process.

The physics establishment scorned the medical doctor's theory. Robert Park, PhD, a professor of physics at the University of Maryland, proclaimed that Mills is not a scientist. Park went further and said science already knows everything about the hydrogen atom.[179]

Park had never tested Mill's work in a laboratory but continued to attack Mills' theory while:

- Penn State University tested Mills' energy process and produced a hundred times more heat than would have been created by *burning* an equivalent amount of hydrogen.

- Mills' company was attracting investors from the electricity industry and the support of energy experts.

- Shelby Brewer, an industrialist and former assistant Secretary of Energy for U.S. president Ronald Reagan, joined BlackLight Power Inc. as a board member.

The head of Reading Energy Company of Philadelphia studied Mills' test results and told Reuters news service that the development is real and that if it keeps going as qualified experts think it will, his work will spark a scientific revolution on the magnitude of that begun by Thomas A. Edison or Albert Einstein.[180]

Park, then public relations officer for the American Physical Society, mainly ignored the replications and validations of the BlackLight Process and instead came up with witty put-downs whenever the company made the news. Park had built a reputation as a debunker; science journalists turned to him for comment when they needed a skeptic. He called hydrinos a "free energy scam" and in 2006 Park said Mills' work is "clearly fraudulent, I don't hesitate to say that."

Mathematics teacher Thomas E. Stolper, who wrote a biography about Mills, said Robert Park is accustomed to shooting at easy targets and grossly misjudged Randell Mills. Park had blasted Mills in a brief section in Park's book *Voodoo Science*, but made at least twelve mistakes about Mills within three pages. Stolper's book corrects each of Park's errors, [Stolper p. 279–281] and documents numerous incidents in which critics such as Park (who have not done the experiments) destroyed BlackLight Power's opportunities.

One serious incident involved interference with the United States Patent and Trademark Office.

BlackLight's patent attorneys had succeeded in getting past what Stolper calls the "poisonous fumes" of the cold fusion controversy, which have closed more doors at the patent office, and on February 15, 2000, the company was granted its first patent for hydrino energy. BlackLight was told that five of its patent claims for hydrino energy had been allowed, and the invention had been assigned a patent number. The company paid the fees and at that point assumed it was safe to celebrate. The intellectual property protection conferred by a patent is considered crucial to attracting investors, and now it was essentially ensured.

It was practically unheard of that the patent office would withdraw a patent from issue at that point. BlackLight's patent was about to be published in the *Official Gazette* when a phone call informed an official at the patent office that the hydrino theory is dubious science, contrary to the known laws of physics and chemistry, and that the patent raised issues similar to "questionable technologies like cold fusion and perpetual motion"[181]. Stolper in his biography tells the convoluted story of who put pressure on whom in this incident, indicating that Park may have been involved behind the scenes since he had bragged about having a "Deep Throat" connection in the patent office. The outcome was that the revoking of their patent applications sabotaged the company's plans for going on the stock market at that time to raise capital.

Poisoned attitudes toward BlackLight reached across the ocean. A

European Space Agency physicist wrote what Stolper describes as a badly flawed analysis of Mills' work and in an email called Mills' work a hoax. That one person's hostile attitude caused the space agency and some companies to end their contact with BlackLight. The journal[182] that published the flawed analysis refused to publish Mills' lengthy response. The editor said it would require considerable persuasion before even a succinct rebuttal would be published.

Quantum physics defrocked?

Meanwhile, Randell Mills continued his work. When and if his theory is widely accepted, Mills said, "I think it's going to be a case of 'The Emperor Wears No Clothes'."[183] In an amiable tone he mocked quantum physics. "There were some problems in the early 1900s in which people said 'Whoops, the photon seems like it's a packet, and the electron seems to have this wave character, so we'll just invent this wave physics that has nothing to do with things we can measure directly.'" Mills added that the physics degenerated into a theory that acted as a patch, and then took on a life of its own. After the prevailing mindset finally changes, he predicted, people will wonder why the illogical errors persisted.

National Aeronautics and Space Administration (NASA) scientists studied the energy production from a Mills cell, and left the door open to an alternate explanation that would replace "hydrinos." So did Mills' plasma physicist colleague in Germany, professor Johannes Conrads.[184] The professor was able to produce remarkably high energy from a Mills cell.

Novel materials

BlackLight Power had for years been focusing on producing new and useful materials, and studying the energy reaction as a byproduct. The company's scientists methodically designed experiments and learned properties of the unusual leftovers from the hydrino reaction. Some of the materials looked ideal for making batteries.

One of the unique materials appeared to be a plastic that is strongly magnetic and contains no carbon, with the potential to improve products ranging from semiconductors to airplanes. Another was an extremely strong anticorrosive coating that apparently could rustproof ships.

BlackLight Power relocated into a 53,000 square foot former Lockheed Martin aircraft hangar in Princeton, New Jersey. After discovering about a hundred new compounds, the company bought a 150-ton vacuum chamber—looking like an oversize thermos bottle—for making larger quantities of some of the new materials.

Environmentally benign

The company's plans for its closed system—a power converter making its own supply of hydrogen from water—would also work on ocean water, Mills said.[185] It will be safe; if the unit were to be punctured, the low-pressured process would stop because it requires a high vacuum.

The process recycles both hydrogen and catalyst. Hydrogen goes into the cell, then the hydrogen molecules are broken into atoms that react with a gaseous catalyst which removes energy from hydrogen without nuclear radiation and creates three items: plasma, power and new chemicals. The new chemicals with extraordinary properties collect in the reaction vessel. Once a year the vessel is taken out and swapped for another one. The chemicals would be retrieved and the vessel put back into service again for another year of having excess electricity to sell to the grid or use locally.

Forging ahead

Mills hoped that a demonstration would cause people to take notice. "And I think then people are going to overlook the arguments and bickering about quantum mechanics, and move ahead. I think things will accelerate pretty quickly after that." Eight years later the prototype is ready and tested. The company's trademarked motto is "Greater Than Fire."

Their research and development timeline could have been shorter if opponents had not interfered with the patenting and therefore caused funding challenges for Mills.

Nevertheless, the news looks positive. Shelby Brewer, BlackLight Power Inc. board member and former Assistant Secretary of Energy for U.S. president Ronald Reagan, says, "The BlackLight Process gives rise to a prospective inexhaustible, economical, environmentally friendly energy source."

Randell Mills takes pride in having had no funding from the government; his company is totally privately funded unlike what Mills calls "Big Science".

We wonder if keepers of the old physics paradigm are themselves in danger of becoming "fringe". Pioneers whom the gate-keepers ignored, loudly ridiculed or hampered have quietly progressed, apparently by drawing on inner resources that grow stronger each time the pioneer survives an attack. We expect that the younger generation will find so much inspiration, excitement and practicality in the emerging science paradigm that they will just pass by the limited thinking of arrogant minds. We also marvel at what one person can accomplish if grounded in noble motives.

Some of the innovators in the next chapter care very little what Big Science has to say about what they're doing. In discussion groups on the Internet they are sharing tips on how to liberate energy from within the atoms of water.

14

Singing Water

During the 1950s Phil Stone, a retired Florida college physics professor, had a patent for a device to run an engine on water. The U.S. Government then unfairly classified his patent, and this prevented him from developing his device.

—*GARY VESPERMAN, engineer*[186]

All is not known about water.

—*EUGENE MALLOVE, PhD*[187]

To claim that numerous recently discovered methods of super-efficient water decomposition. . . are impossible because Faraday did not discover them, is like saying that space travel is not possible because the Wright brothers did not discover it.

—*JAMES ROBEY, author*[188]

On our blue planet with its oceans and lakes, many inventive minds are seeking a super-efficient water-splitting technique to extract cheap hydrogen that can be used as clean fuel to replace gasoline.

Their critics say that what some of the innovators are doing is impossible; it violates a venerable law of science. In the early nineteenth century British scientist Michael Faraday figured out that the amount of electrical current zapped into an electrolyzer (an apparatus that uses electricity to break apart water) is proportional to the output of hydrogen and oxygen gases. Faraday's Law means that to produce one cubic meter of hydrogen, nearly four kilowatt hours of electricity is needed.[189] However, Faraday used what the innovators are figuratively calling "brute force electricity" in contrast with their more gentler currents.

The water-as-fuel concepts coming from the international energy underground are not part of the establishment's well-advertised Hydrogen Era that would feature:

- massive pipeline and tank infrastructure to transport and store compressed hydrogen

- construction of more nuclear power plants to provide electricity for electrolyzing hydrogen out of water; and

- continued reliance on fossil-fuel natural gas (methane) as a source of hydrogen

- extensive use of expensive fuel cell technology to combine hydrogen and oxygen to produce electricity.

Instead, some prototype systems designed by the outsiders use existing internal combustion engines to burn the hydrogen that is generated onboard. Some inventions run on ordinary tap water, others on salt water. In the new energy approach hydrogen gas is made "on demand," only in amounts needed by an engine in the moment. Publicly advertised or secretive, these revolutionary projects are springing up in all regions of the globe, from the U.S. to Japan to the Philippines to Australia and elsewhere. Anyone can search for "water as fuel" on the internet and watch hours of intriguing on-line videos that describe these developments.

Puharich's water-fueled camper

Subjecting the water molecule to specific frequencies as a splitting technique was one basis for inventions of two larger-than-life personalities in the recent history of the new energy field, Americans Stanley Meyer and Dr. Andrija Puharich. A Canadian company is continuing their legacy.[190]

Andrija "Henry" Puharich (1918–1995) was a medical doctor who also had a PhD in physics, thirty patents and many papers and books to his credit. His work in science was mostly self-financed, spiced with an independent attitude that made him a maverick.[191]

After analyzing Nikola Tesla's writings about electrical resonance, Puharich discovered an easier and perhaps gentler way to break apart the water molecule—by rhythmic shaking at its natural frequency of vibration. Puharich patented his method of splitting water molecules.[192] Dr. Andrew Michrowski of Ottawa describes the method. "Complex electrical waveforms form resonant frequencies of tetrahedral water molecules. These waveforms resonate water molecules and shatter them, thereby liberating hydrogen and oxygen."[193]

Puharich was an adventurous scientist, and Michrowski and other friends tell about cross-continent trips Puharich took in his mobile home powered by hydrogen split out from water as fuel. A Puharich legend depicts him melting snow on a Mexican mountain pass to fill his tank. Michrowski reported that the inventor drove several thousand kilometers with the water fueled camper, on one occasion using water found on salt flats to keep the engine going.

Onboard batteries provided the electricity for the water-splitting. It is claimed that Puharich's yielded much more energy from the combusted hydrogen than was required as electrical input energy to generate the hydrogen (an example of "overunity"), and therefore had advantages over a pure electric vehicle, such as a vastly increased mileage range. A physicist friend reported that "the AC electrolysis resonator in Puharich's device is resonant; hence the process has greater efficiency than expected had the system involved a near-equilibrium linear process"[194]. In other words, Puharich's electrolytic cell was not a conventional closed system, and it could tap into energy from the surrounding environment.

Like many brilliant inventors, Puharich's strengths did not lie in carrying an invention forward to the point where it became a product on the marketplace. Instead, he turned to other intellectual challenges, contributing to human understanding of the nature of consciousness.[195]

Stan Meyer's water fuel cell

Stanley Meyer did not get a chance to go on to other challenges after patenting his invention. Although he looked healthy and had the physique of a former football player, he died suddenly in mid-life. The story that sped through the new energy community was that he collapsed in the parking lot outside a restaurant, claiming to have been poisoned. He was apparently celebrating having finally secured funding to build a production facility for his Water Fuel Cell (WFC). We were told that the coroner's report cited natural causes for his death.

The first time Jeane Manning interviewed Meyer was a decade before, when he spoke at a conference in Switzerland.[196] The hearty-voiced American towered over most of the crowd. Translators couldn't keep up with him, because he always talked rapidly as if time were running out. He told of meetings with leaders of different countries who need an answer to the energy problem.

During his speech a video showed Meyer driving a dune buggy for a public demonstration.[197] The buggy had a Volkswagen engine running on a water fuel cell powered by batteries. Meyer claimed that the vehicle would cover twenty-five miles per liter (95 miles per gallon) of water. As with Puharich's, his invention was quite different than standard electrolysis. It was based partly on a view of the water molecule as held together by bonds that can be tickled apart by a small input of AC electricity at frequencies resonant with the "electrovalent bonding" of the molecule.[198]

Meyer began his technical career in electronics and electrodynamics and worked for a time at the Battelle Institute in Ohio. Then he owned an auto parts business and made what he called a fortune which he later used to finance research. He told Jeane that he had had a fleet of transport trucks during the 1970s and realized their vulnerability to an oil embargo from the Middle East. At a meeting with other business leaders during the threat of oil and gas cutoffs, he said, "I saw some of the most powerful industrial leaders of the state of Ohio popping pills; I thought they were going to have a heart attack. They said to us 'without energy we can't make a product; if we don't have any profit we can't pay our bills'."

That was when he decided to solve the technical problems of using hydrogen from water as a fuel source. The first challenge was to release the hydrogen gas from water economically. The electrical polarization method he developed is opposite to electrode-destroying standard electrolysis. In contrast, the WFC used ordinary water in a non-chemical environment more comparable to an atomic particle generator than to electrolysis. The WFC was a composite of inventions. A series of patents claimed to produce the gas on demand and make it safer than natural gas. Water was exposed to a particular pulsed voltage intended to electrically induce the water molecule to release the bonds between its atoms, then the released hydrogen was prevented from re-forming into water molecules until ignition. "An avalanche effect occurs which continues to release thermal explosive energy in a very safe way."

Despite the elegance of his method, Meyer became a controversial figure because of selling business dealerships. A court of law condemned his business practices. Some scientists also found him difficult to deal with. A strong personality with outspoken views, he caused some professionals in his audiences to distance themselves when he mixed Biblical quotations in with technical presentations. He also alienated members of the research community who visited him, by disappointing their expectations of seeing working hardware.

Since Meyer's background was more in business than in the science community, he believed that securing patent protection took priority over sharing of information. As with Floyd Sweet, a scarcity-based viewpoint lingers even in those who have seen abundance of energy. Meyer also believed that he had a responsibility to keep the technology for his own country and its allies rather than sharing it with all nations.

Fears

"If we went ahead and got the patent on the electrical polarization process and did not get it on resonant action—tuning it in to the dialectric properties of water and their resonance—and if we did not secure the patents on the voltage intensifier circuit or the electron-extraction circuit, or the quenching-circuit technology, then it was possible that a hostile foreign entity could have filed blocking patents," Meyer said.

Did his long-standing fear of "foreign entities" finally attract that hostile presence into his life? We can only speculate. He sometimes mentioned having received death threats and "proferred multi-million dollar inducements."[199] Meyer said he had meetings with both military and government leaders but was not able to release information about any collaborations. As he told it, in a meeting in the Pentagon in the mid 1980s Meyer posed a question for the military officials. "Gentlemen, if you don't have an economy, you're not going to have a government. If you don't have a government, you're not going to have a military. So this is imperative to get this type of technology into the economy. . . ."

Source of energy?

The WFC does not create energy, it only triggers the release of energy that already exists in the hydrogen atom, Meyer said. "When you run the car on water, the byproduct of burning water to release the thermal explosive energy is water mist. When it goes out the exhaust, that water mist could be slightly de-energized. So you simply let the water be exposed, evaporate, go back into the atmosphere, be exposed to the sun's energy. The molecule will absorb that photon energy and the molecule is simply re-energized again."

"When the water comes back in the form of rain, put it in a car. . . .photon energy is being released inside to drive the car. All we're doing is using solar energy to use water as a source of fuel."

Admiral Sir Anthony Griffin, Controller of the British Royal Navy for five years before retirement, was a supporter of Meyer before Griffin's own sudden death. Griffin wrote a formal report, *Water As Fuel*, on the WFC and on Meyer's newer development "Hyperdrive".[200] Stan Meyer wanted to see vehicle engines run on hydrogen and repurify the air. He was passionately concerned about the depletion of oxygen above highways.

On his visit to Stan Meyer with a colleague, the late Eugene Mallove, PhD saw "a rather amazing demonstration of something I can hardly imagine is standard electrolysis. There wasn't enough thickness of wire to hold enough current to do what he did, right in front of our eyes." Mallove said they left Meyer's workshop with "a very impressed kind of stunned reaction," but could not know from that encounter whether Meyer could perform the wonders that he said he could."[201] The fiery Mallove and Meyer clashed angrily after Meyer did not allow the visitors to do measurements on the system.

Meyer's patents eventually expired. Researchers who are building variations of his system today believe that Meyer did indeed use water as fuel, but the secrets of his complex method are still being uncovered. There are also other methods to solve the clean energy supply problem.

Global water fuel efforts

Filipino engineer Daniel Dingel began working on a water-car idea in 1969 and claims to have converted more than 100 vehicles to run on a system that uses sea water. However he says politics and fears of lost fuel-tax income prevented his government, from Ferdinand Marcos' era to today, from helping him to patent and take the water-car to market. The World Bank and International Monetary Fund can tell the Philippines, a debtor nation, what they can or cannot do in the marketplace, he says.[202]

In a way, Dingel's invention is the opposite of the process used in standard fuel cell car developments (not to be confused with Stan Meyer's Water Fuel Cell). As with conventional electrolysis and the methods invented by Puharich and Meyer, Dingel's method uses an electrical input to release hydrogen from water, and the hydrogen is then burned in place of gasoline in an internal combustion engine. The explosive combustion of hydrogen with oxygen in the engine cylinders moves the pistons to power the vehicle.

In contrast, a fuel cell vehicle is actually a type of electric vehicle. In a standard fuel cell, hydrogen is combined with oxygen electrochemically to produce electricity, and that electricity goes to electric motors that power the vehicle. So a fuel cell simply uses the reverse reaction of electrolysis.

Conventional physics says it is impossible to get enough electrical power out of a car battery to do what Dingle claims to be doing. Skeptics point to slight traces of carbon found in his car's exhaust suggesting that he is still using some hydrocarbon fuel, although that could have arrived in the air intake from smoggy sooty city air.

Videos on the Internet show Dingle running a car with a small hydrogen reactor that he invented hooked up to its engine. His "concept car" is a red Toyota Corolla.[203]

Cheap hydrogen, from Russia

Skeptics would have a difficult task if they challenge the scientific credentials of Russian physicist Philip M. Kanarev. His recent job titles include professor and head of Theoretical Mechanics at Kuban State Agrarian University.

He and colleagues found that when water molecules are shaken apart by using electric pulses at frequencies that matched their own natural resonant frequencies, the molecules' "valence electrons" absorb energy from the vacuum of space and as a result the process can output up to ten times more hydrogen than ordinary electrolysis.[204] In keeping with the scientific method, Kanarev has both theoretical and experimental results to show as well as the mathematics of his successful "plasma electrolysis." His country has little funding for such research, however.

Dr. Kanarev's low-budget breakthrough hydrogen experiment (photo credit: Juha Hartikka, Finland).

Around the world with water-as-fuel

- Zlatco "Shad" Loncar of Croatia uses a vortex system to work with water as a medium to transform energy. Receiving more energy than invested for the ionization means that some other energy helped ionize the water.

- One American company has an over-unity (a third more heat output than equivalent electrical input) commercial boiler on the market. The ShockWave Power™ Reactor harnesses the "water hammer" effect, called cavitation, to heat liquids in seconds without flame.[205]

- A duo of engineers has novel waste-to-energy process that converts sewage and industrial liquid waste to a gaseous fuel that could run a fuel cell. It's still dealing with carbon, but provides a solution to waste issues *and* a fuel alternative to extraction of oil and coal.[206]

- Many people in North America have seen a news item about an amateur inventor burning seawater. John Kanzius of Pennsylvania invented a machine intended to treat cancer tumors with radio waves. While experimenting, he targeted a test tube containing salt water. Instead of boiling off as expected, salt water became a gas that he could burn. Scientists who studied the process decided the gas is probably hydrogen. Will the first commercial water-car be advertised as Powered By Radio Waves?

Do-it-yourself hydrogen use

- In Phoenix, Arizona, university instructor Roy McAlister formed the American Hydrogen Association to provide liberating information to do-it-yourselfers wanting to convert their vehicles or lawnmowers to run on hydrogen. A.H.A. teaches about zapping hydrogen from water, using Sterling engines to run on the heat from solar concentrators, getting hydrogen from plant waste and how to convert existing engines to run on hydrogen.[207]

The oxygen issue

- Professor Ruggero Santilli of Florida,[208] has a new technology that uses an electric arc (high voltage, 7000 degree Fahrenheit spark) submerged in liquid waste to generate magnetically-polarized gases that burn. His MagneGas™ is an alternative fuel that puts out more oxygen when burned than it consumes. Florida's Dept. of Environmental Protection recognizes his PlasmaArcFlow™ (a process that converts plasma into combustible gas) for cleaning up sites that are petroleum-contaminated.

A gas with many names

Today's water-fuel inventors look back at an international history. For instance, a Bulgarian emigrant to Australia, Yull Brown, hoisted a car up in his garage and ran the engine on water that had turned into "oxyhydrogen" gas that later was given the popular misnomer "Brown's Gas."[209] Brown moved to North America and then to China in efforts to commercialize a welder that runs on the gas.

More than a decade earlier, physicist William Rhodes from Arizona had patented a process for creating the gas. He was not amused when he heard it called Brown's Gas. Instead of correcting the name and calling it Rhodes' Gas, today's experimenters prefer to say *electrolytic gas, e-gas* or *hydroxy*. They inject the hydrogen/oxygen mixture into engines as a booster to get more mileage from gasoline, or make variations of Rhodes' multicell[210] design—as compact in size as a car battery—in attempts to run vehicles on water.

One of those experimenters uses the name Electric Ride for anonymity when he is online in discussion groups. His story has elements in common with other hydroxy experimenters. After serving in the U.S. Navy on a submarine whose oxygen supply came from an electrolyzer he was ready to pursue his passion, especially when he became appalled at the cost of fuel. The pain level rose when he thought about relatives fighting in an oil war in Iraq. Now for hours every day he shares information with other would-be builders of a fuel booster and with their feedback information he built a unit that increases gas mileage by 20 per cent. It's encased in a four inch diameter one foot long plastic tube and can be scaled for lawn-mowers or cars. He has one installed on his motorcycle. It cost only fifty dollars for the parts, but the meticulous spacing of stainless steel plates takes a lot of time. When he was interviewed by radio host James Robey[211] they agreed that high schools and colleges should have classes on water as fuel. "Every time I hop on my bike I'm disassociating fuel," he said with apparent satisfaction. "You have to take power in your own hands."

15

Magnets in Your Future

I think it is possible to utilize magnetism as an energy-source. But we science idiots cannot do that; this has to come from the outside.

—WERNER HEISENBERG, *Nobel laureate*

Locked magnetic electron vortex spin can power all of our energy requirements forever.

—ROBERT H. CALLOWAY, *inventor*[212]

Where can you find something continually tapping into zero point energy or the ether? Inventors of innovative motors and generators using novel magnet arrays look to the subatomic level of matter to find that happening. Let's follow their thinking for a moment.

All magnetism arises from the movement of electrical charge. In an *electromagnet*, a coil of wire that carries electrical current induces a magnetic field. A *permanent magnet* also gets its magnetic effect from the movement of electric charge. A permanent magnet does not need a current-carrying wire, however. Its magnetism comes from a *spin of electrons* in the material of which the magnet is made.[213] The more aligned with each other the magnetic domains in the material are, the stronger the magnetic pull.

And from what source does an electron get the power to keep spinning perpetually? A number of peer reviewed physics papers[214] point to the *zero point energy of the quantum vacuum* as the power source that prevents electrons from ever wobbling to a stop or falling into the atom's nucleus.

Where else could we find the cause of unceasing spin throughout the universe? Nassim Haramein pursued the question. Particle physicists didn't want to talk about "causation" and he was finally told that the origin of spin was the Big Bang. Maybe that reply would make sense in a frictionless universe, he reasoned, but even in outer space atoms occasionally collide and would have slowed down over eons. Haramein followed logic, observed nature, worked out the underlying geometry of space and concluded that the perpetual twistings of spacetime are powerful enough to cause the relentless spinning.[215]

If the incredibly powerful invisible background is the ultimate origin of spin and therefore of magnetism, then a self-running magnetic motor can't be accused of violating any laws of physics. The invention is not pulling energy out of nowhere as the Irish company Steorn speculated when first asked to explain their own anomalous motor. As would be expected, Steorn was immediately branded with "perpetual motion . . . violating the First Law of Thermodynamics."

It is impossible to give credit to all the advances in magnetic generators coming from researchers around the world. Part of that contribution comes from lone inventors who poured their lives and thousands of dollars into the quest for energy independence. Their efforts gathered momentum in the later part of the twentieth century when extremely powerful permanent magnets became available. One magnet of a certain grade made with the material neodynium and weighing only seventeen ounces has a pulling force of 640 pounds. And that's not the strongest grade. These magnets can break bones if your fingers get caught when two such magnets slam together.

The attracting as well as repelling force of these magnets can certainly do work. Their power declines so slightly that it means little—about one per cent every ten years according to some estimates. The work the innovators want done is converting mechanical power into electricity. And the challenge is not in finding how to allow the magnets to pull themselves together repeatedly, but how to pull them apart repeatedly. If it takes as much work to jerk them apart as the amount of work magnets performed by slamming together, no energy is gained for turning an electric generator.

The late Bill Muller whom you met in chapter one created a way to make a rotor spin easily by cleverly counter-balancing the magnets' reactions to each other in space and in time. He did this by a combination of mechanics, timing, circuitry and material science. "Perpetual motion is stored in the magnets. The Magnets are the Motor," he often said. A team in Texas has replicated Muller's generator and as this is written they are testing a prototype to see if it has more output than the apparent input.[216]

One of today's inventors finding a way around that dilemma of how to get an energy gain is Alan Francoeur of British Columbia. His designs for innovative generators include an "Interference Disc Generator." Instead of rotating magnets on a wheel, it has a spinning iron shield whose spokes alternately interfere with magnetic fields and then allow the field to permeate wire coils and generate electricity. One benefit of the mechanical design is that it needs no complicated or fragile electronics.

Francoeur's passion for such concepts started when he worked at Giant Mine in northern Canada in the mid 1980s as a teenaged mill operator. The mine's gigantic and old electrical generators were daily exposed to his eyes and imagination. During his break time, small magnets and pieces of tin or wire were in his hands or on the table in front of him as he contemplated the shielding of magnetic fields possible with a piece of metal. When he returned to Calgary in 1986 he built generator models and caught the attention of a medical doctor[217] who helped him buy magnets and machining. Later after the Internet sped up the sharing of inventive ideas, a Russian scientist patented one of Francoeur's magnet motor concepts.

Alan Francoeur and his AC DC three phase dynamo.

Recently Francoeur phoned co-author Manning excitedly. He felt vindicated, having come across documents about a radio pioneer whose large alternating current electrical generator, with similarities to Francoeur's smaller machine, was well known and successful in the era of vacuum tubes when high-amperage radio frequency currents were needed. A working model exists in Sweden. From descriptions of the Alexanderson alternator, Francoeur deduced that the larger dynamo lacked one benefit; Francoeur's rotor disc is magnetically balanced for easy rotation.[218]

Although frustrated at this time by lack of money to buy all the time, materials and machining needed for experiments, he is fortunate in that his wife Jan is supportive of the obsession. In fact she joins him at the kitchen table to help with the laborious winding of electrical coils.

Not every innovator is working outside the mainstream. Joseph Flynn's company Flynn Research Inc. in Greenwood, Missouri, is a registered government contractor and performs research and development tasks for the military, according to the company's website. At a conference attended by leading aerospace researchers, Flynn presented a paper on its "Parallel Path Magnetic Technology for High Efficiency Power Generators and Motor Drives."[219] Flynn has access to sophisticated testing equipment as well as specialists.

The output from a motor more than triples when flux fields within a core are manipulated according to the Flynn method, and a patent points to possible solid state generator applications. "All motors henceforth should employ this technology," states the New Energy Congress (NEC) listing. The New Energy Congress is a largely volunteer association that reviews the most promising claims to up-and-coming clean, renewable, affordable, reliable energy technologies and posts a Top 100 online. The Flynn magnet powered motors and generators are in the NEC's database of the top one hundred new energy technologies.

However Flynn Research has apparently been reluctant to openly share the wealth of knowledge it has collected. This close-to-the-vest corporate model of operating contrasts with the open sharing of information among independent researchers such as Francoeur, who has posted prolifically online.

Generator flies away

Outside of North America, an outstanding figure in new energy lore is John Searl of England, whom Manning first heard about through the NASA consultant Rolf Schaffranke (chapter one) and then met for the first time in Switzerland in 1989.

Searl's experiments were unprecedented. In an effort to develop a generator that runs itself he made a magnet-powered saucer shaped craft that he said produced its own gravity field, repelling the earth's field and rising into the sky. A few photographs exist to show that test flights began in the 1950s. Newspaper photos [220] taken between 1973 and 1975 show Searl pointing to the disk rising above a field and explaining to a reporter how it flew powered by an arrangement of a large ring magnet with rolling magnets in channels around the outer edge. The experiments were guided by information he had received in childhood dreams. Searl worked for a British utility board and could use its laboratory with the equipment and tools he needed.

His wife became unbalanced by the public attention given to her husband and colluded with the power company to get Searl jailed on charges of stealing electricity from the power grid. He had indeed been running their house on a Searl Effect Generator for years. While he was in prison she presided over four days of the burning of his books and records. When he returned home he found the house burned, his equipment and models missing and learned that the family had moved to an undisclosed address. Homeless and despondent, he walked into the countryside to find a friend to take him in. His health deteriorated.[221]

For years Searl was spoken of almost in the past tense, since researchers knew he had suffered devastating personal losses as a result of his experiments. Then at the Swiss Association for Free Energy conference in Einsiedeln in 1989, one evening a side room filled to overflowing after an announcement that the legendary John R.R. Searl would give a talk. The eyes of onlookers moistened as he shared the pain of those years, but when he vowed to carry on his work the doubters murmured that "he won't remember how he made the magnets . . . too old, been ill . . ." Others believed that he would make a comeback.

He did not give up on getting funding and progressed slowly with various helpers. Because he spoke in terms of the information received in his dreams, which he refers to as the Law of the Squares, he was often treated dismissively by people who otherwise may have helped.

Two Russian scientists, Vladimir Roschin and Sergi Godin, saw a videotape of a talk Searl gave in Germany in 1990 and decided to try to replicate his work. They built a unit that rotated an electrical generator with an output of six kilowatts (enough to power two average households) for at least fifteen minutes without consuming any electrical energy according to one Russian scientist who knew the duo. Amazingly, the effect of gravity on the rotor decreased about 35 per cent.[222]

Effects were dramatic and unexpected. After the starter motor accelerated the rotor speed to 550 revolutions per minute, the starter motor was no longer needed; the system was self-running and became a generator of power. When they turned the lights off in the room, an ionization cloud—pink and blue light—was seen in a donut shape around the machine with a wavy pattern corresponding to the surface of the rollers around the outside edge. The scientists' instruments also detected concentric rings of vertical "magnetic walls" around the energy converter. And the temperature near the converter dropped 6–8 degrees Celsius. When one of the scientists went upstairs to a second floor room above the laboratory, he measured a similar temperature pattern whose changes corresponded to the magnetic walls of differing magnetic densities. Steel-reinforced concrete blocks separated the room from the laboratory below.[223] When the Russian scientists showed up at a conference in 2001, they had lost that laboratory and the prototypes due to some problems with an investor. Thomas Valone became aware of their experiments, found an American investor to fund them and later helped them write a patent on an improved model.

As this is written, John Searl's team has built a model to investigate how electricity is liberated as the rollers make their rotation. A colleague of ours is planning to fund the Searl effort. These few samples of the research being done point out that the science establishment as well as maverick experimenters still have a lot to learn about magnetism, and there is a strong potential for widespread replacement of carbon fuels energy by harnessing the power of magnetism.

In the next chapter we take a fast trip through some of the wide variety of breakthrough energy discoveries. Just as Joseph Flynn may not be happy about having his advancement mentioned in the same chapter as developments that are more controversial, some inventions in the next chapter are widely separated when it comes to the backgrounds or viewpoints of the inventors. However, given the stakes for our planet, isn't the overall goal a united effort to advance those technologies with potential to replace polluting and expensive energy choices?

16

Downpour of Solutions

We expect small wind turbines to become an increasingly common sight. Renewable energy doesn't have to be large scale.

—*MIKE THORNTON, head of Energy Saving Trust Scotland*[224]

Multiple methods for producing vast amounts of energy at extremely low cost have been developed.

—*DR. PETER LINDEMANN, New Energy author*[225]

In discovering fundamental omissions in thermodynamics, I feel like I have discovered a whole continent of possibilities! . . . I feel like I am living in the 1300s; everyone believes the world is flat and I am saying the world is round.

—*KEN RAUEN, New Energy scientist*[226]

So far in *Breakthrough Power* you have been introduced to energy technologies that involve magnetics, hydrogen chemistry, implosion/vortex, sound, electronic circuits, resonance, water, nuclear reactions without dangerous radiation, tidal currents, super-efficient electrolysis and zero point energy devices with no moving parts to wear out.

Breakthroughs in the ingredients for some energy converters, such as speedy electronic switching, powerful permanent magnets, and other new materials, have brought technology closer to an energy revolution. (A further breakthrough needed is in a realm that we, the people, can affect—political will.)

This chapter samples some of the variety of additional ways that humankind can power its civilization affordably and fairly. A rapidly increasing number of promising, and often proven, energy inventions continually surface on the Internet. The New Energy Congress (NEC) website[227] reviews systems spanning the spectrum from conventional such as solar and wind to highly unconventional alternatives. NEC is a private association of experienced new energy investigators formed to review the most promising claims of emerging clean, renewable, affordable, reliable energy technologies. (Sterling Allen representing the NEC was inducted into the new Environmental Hall of Fame[228] in Los Angeles on World Environment Day 2008. Four other recipients were "green energy" companies then listed on the New Energy Congress' Top 100 Clean Energy list.)

Following are some examples of the rainbow of revolutionary clean choices offered to the human family, some more energy-dense (power output per unit of size) than others. When you see wording such as "holds promise", keep in mind that lack of funding for the expensive development, testing and pre-manufacturing stages has prevented many promising discoveries from becoming proven and commercially available products.

Nano-engineering

Moving things around on the molecular level where measurements are in the nano (billionths of a meter) scale, researchers are coming up with breakthroughs in efficient solar collectors and batteries. To get an idea of the miniscule sizes involved in nanotechnology, a virus particle is about 100 nanometers wide. Nanotechnology is a fairly new field and stepped into the media spotlight in the late 1990s after scientists learned to control matter, and make devices, smaller than that virus. When sizes are shrunk that much, materials may demonstrate unique and valuable properties, such as enhanced strength or electrical conductivity. New materials also promise breakthroughs in converting heat directly into electricity.

Electric sunshine

A company called Nanosolar appears to have a breakthrough in photovoltaic solar cells 1/100th the thickness of conventional silicon-wafer cells. An extremely thin light-sensitive film is sprayed or printed onto flexible material that can be rolled out on any surface, so it could be a part of building materials such as roof tiles and windows. Nanosolar says their panels will be the world's lowest-cost solar panel, as low as 99 cents per watt.

A variety of other solar advances are on the NEC Top 100 regularly. If they are mass produced and if oil and nuclear subsidies are diverted, solar power could replace a significantly larger chunk of the oil empire than most people realize.

Waste heat

Nanotechnology breakthroughs make it possible to produce electric power from many different sources of waste heat, ranging from industrial scale processes to our own bodies. Imagine wearing a personal power-jacket that uses body heat to recharge electronic devices.

These "thermoelectric" materials convert heat directly into electricity. Energy now lost as heat during the production of electricity, industrial scale processes and the internal combustion engine could also be harnessed.

Nearly all of the world's electrical power and transportation is generated by heat engines. Giant gas or steam-powered turbines convert heat to mechanical energy which is then converted to electricity. About fifteen trillion watts of this heat is not converted but is instead released into the environment. If even a small fraction of this lost heat could be converted to electricity, its impact on the energy situation would be enormous.

Thermoelectric materials could be used to capture much of the low-grade waste heat now being lost throughout industry and convert it into electricity. This would result in massive savings on fuel and carbon dioxide emissions. The challenge for scientists and engineers has been to make thermoelectric materials that are efficient.

Waste heat from engines could be converted into electricity to recharge the batteries of hybrid vehicles and extend their mileage range. Or the batteries could dump their stored electricity into the grid once the vehicle is back in the owner's garage.

Saved on silicon

Researchers at the U.S. Department of Energy's Lawrence Berkeley National Laboratory have shown that it's possible to significantly enhance thermoelectric energy efficiency at room temperature in specially processed rough silicon nanowires. The technology could also be used in refrigerators and air conditioners.[229] Thermoelectric modules could convert heat from car exhaust into supplemental power for a hybrid electric vehicle or to run a car radio and air conditioner. When scaled up, the modules could co-generate power with gas or steam turbines.[230]

Pioneering heat into electricity

Peter Hagelstein, PhD of Eneco Inc. was first to patent a "Thermal Diode for Energy Conversion."[231] His diodes would convert heat into electricity so efficiently that with only a ten degrees Celsius temperature difference, a pool of water six meters on a side could supply the electricity for a house. Hagelstein also suggested their use as efficiency boosters for electric or hybrid cars.[232] Eneco described its product as a new type of semiconductor with five times the energy density of lithium-ion batteries, and scaleable from very small to very large applications without losing efficiency or cost effectiveness. The company had contracts with the military.

Nitrogen as fuel

Ed Sines is a former industrial scientist now working on a small scale system for a potentially big breakthrough in generating electricity. He's already made a breakthrough in thin films. A nuclear physicist who supports Sines' work says the invention may be able to function on the megawatt scale, to replace current electrical power plants. It creates a superconductive state from which to gather energy from the background environment. The only waste products are heat and nitrogen gas, found in the air naturally.

This process for generating electricity would include pulsing of a laser into magnetic fields to change the conductivity of a liquid nitrogen cooled superconducting thin film on tiny quartz tubes. Liquid nitrogen is cheap and readily available. If the invention succeeds, as a number of energy researchers believe it can, there will be no need for fossil fuels to generate electrical energy. A prototype the size of man's hand is expected to be powerful enough to run household appliances. It would be safer for powering vehicles; in an accident liquid nitrogen would help put out fires, in contrast to today's toxic, combustible hydrocarbon fuels and explosive batteries.

After his more than 15 years of pioneering research on the system Eddie Sines believes only a modest amount on funding (less than $250,000) is required to launch his prototype. That isn't asking for much, compared to the potential benefits of his invention.

Hydrogen fuel breakthrough

In California, an inventor[233] with a background of twenty-five years in chemical physics has done what has never been done before—he has combined a gas and a solid to make a carbon-free room temperature liquid that can be used as a hydrogen fuel. A car engine ran on a batch of the substance in road tests in 2008. The process requires hydrogen, but the inventor is doing research and development of a second invention, synergistic with his other invention, that would be a novel source for "cheap, green" electricity for electrolyzing water to get pure hydrogen. That second breakthrough has not yet been built. It can be called a nonconventional heat pump, yet it involves nano-engineering technology. Its heat pump characteristics have the potential to revolutionize air conditioning systems, refrigeration, furnaces and heaters.

Wind breakthrough

Colleagues in Europe have been telling us about a revolutionary approach to wind power that would be quieter, more versatile and with greater flexibility to tap wind from any direction compared to conventional wind turbines. Electronics built into the system control multiple electricity generators, so the windmill would harvest power from a wide range of wind speeds.

The European company is not seeking publicity at the moment, but elsewhere other improved harvesters of wind are on the open market. If your mental picture is of either the conventional creaking Dutch windmill or a monstrous wind farm, consider the new designs for rooftop generators. There have been breakthroughs in efficiency through improved magnetic generators and in quieter noise levels. The New Energy Congress website points to some of the better windpower products.

Plasma-based energy breakthroughs

In addition to solid, liquid and gaseous forms, matter can exist in a very expansive fourth state called plasma. It's created when electrons are stripped out of a gas, normally by heat or high voltage, resulting in a mass of charged particles that are electrically conductive. Plasma occurs naturally in interstellar space and in the atmospheres of stars. A lit fluorescent or neon light bulb is an example of a highly contained plasma.

Moray King[234] searched peer-reviewed science papers to find out how to activate zero point energy in useful machines. King's role in the new energy scene includes giving practical tips to experimenters, such as "first create a glow plasma or corona (a corona is the glowing region of ionized air around an electrical conductor that has reached a critical point electrically) then abruptly pulse that plasma."

We are including that glimpse into frontier science only to give the non-technical reader assurance that many of the new energy scientists have evolved beyond secrecy and are dedicated to sharing how-to knowledge. They want the inventions to be perfected and the world to benefit.

Energy from pulsing plasma

In Canada, Paulo Correa PhD and his wife Alexandra have made breakthrough in plasma physics as well as validating some of the concepts of Tesla and Wilhelm Reich. Alexandra is an expert glassblower who makes the special vacuum tubes that are central to their plasma device. Their experiments show it is possible to release more energy from gaseous ionized metals in a vacuum than the amount of measurable electrical energy put into the system. Eventually they learned how to get the plasma to pulse on its own.[235] Other physicists had not yet learned a method to excite plasma so that it results in self-sustaining oscillations in the plasma discharge. The Correas made a breakthrough for extracting free energy, but the world of official science generally ignored it.

Another phase of their work showed that the energy involved was mass-free, and that they worked with non-electric reactions of that "aether".

The Correas' aether motor was an electrical generating system that could deliver electrical power without any external power input save its connections to two orgone accumulator (see chapter 8) boxes or a ground pipe. A witness describes a demonstration.[236] "Since the device moved a motor and drove a circuitry, it had to consume some power; this appears to have been provided by the environment. The event occurred with incredible calm—no explosion, no noise even, no sudden heat, no bright light, just the quiet pulsation of a discharge tube and a quiet turning of a small rotor."

The late Eugene Mallove witnessed successful experiments at Correas' laboratory but was unable to persuade the couple to follow his suggestions about commercializing a product.[237]

Plasma focus fusion

Astrophysicist Eric Lerner has been pursuing "focus fusion" for many years. His fusion method is small scale and simpler than the government-funded Tokamak hot fusion reactor, doesn't produce radioactive waste, and holds promise for directly injecting electricity into the power grid without a need for turbines. He estimates that once developed, a commercial focus fusion generator capable of several megawatts electrical output (enough power for 1000 homes) would only cost around $300,000 to make.

Focus fusion is very different from any other fusion approach. Eric Lerner and colleagues use the plasma's instability instead of trying to change its nature. The process uses a laser to help generate the intense heat required to fuse the atoms in the fuel, a hydrogen-boron compound. There is no radioactive waste disposal needed. Lerner was seeking funding for a two million dollar experiment to prove the principle.

Chukanov's ball lightning

A thermodynamics specialist originally from Bulgaria, Kiril Chukanov PhD, harnesses the power of artificially generated ball lightning to extract zero point energy. He would have benefited financially by continuing as a professor, but since 1988 has been trying to bring abundant electrical power to humankind despite limited funding.

For a short period of time he worked on cold fusion at the University of Utah and has written books describing methods for producing electricity from zero point energy. He learned to create ball lightning in a quartz sphere within an industrial microwave oven in his laboratory and found the phenomenon has unusual electrical features and enormous possibilities for generating useable energy. Chukanov's methods are based on conceptual leaps in quantum mechanics. Ball lightning is a "macro quantum unit" and not a plasma in his view. Although he creates bright spheres ranging in color from white to blue, "I don't claim that the ball lightning nucleus has a temperature of 10,000 degrees K." and his ideas differ from today's accepted scientific worldview.

At the time of our witnessing his dramatic demonstration in 2004, Chukanov was seeking investors to fund further engineering development of his process.

Laura Chukanov with ball lightning generator.

Charge clusters

Ken Shoulders pioneered at least twice; he developed much of the technology for microcircuits (he invented integrated circuits.) In recent years he discovered how to make tiny clusters of electrons like miniature ball lightning bunched together in a way that conventional science thought impossible. Shoulders named these clusters Exotic Vacuum Objects, EVOs, and began working toward direct electrical-output generating systems using them. They propel themselves, and the propulsion energy was proven to be many times more than the electrical energy that Shoulders put into the experiment.

The source of all that extra power isn't proven, but Shoulders theorized that it may come from zero point energies. He speculates that zero point energy comes in whenever the motion of the charges is resisted. The self-propulsion is tied to the EVOs self-organized state which also keeps the large number of electrons incredibly close to each other. The microscopic energy anomaly creates such powerful effects that it destroys materials used as targets, but Shoulders has figured out how the EVOs could possibly be contained and emerge as a more controllable form of miniature ball lightning. As with other frontier science by independent geniuses, proper funding could carry this research project forward.[238]

Water arc explosions

How could cold fog droplets and electrical arcs combine to give us a source of power? A self-assured white haired scientist, Peter Graneau, PhD, did not know he would be led in the direction of finding the answer to that when he began a series of experiments in the 1980 at Massachusetts Institute of Technology. Those experiments led to his learning that anomalous energy and force can come from exploding water. It is outside the realm of standard science.

His research team frees excess energy from pure water by applying a high-voltage electrical spark (like a short piece of lightning), to a thimbleful of water and making the water explode with a force that cuts neat holes in quarter-inch-thick aluminum plate. More kinetic (motion) energy comes out of the water explosion than was put in through the electricity. Strangely the water does not get hot. The next step in this research is to learn the best way to harness the violent kinetic energy to make useful electricity. As did Stanley Meyer, Graneau theorizes that water stores energy from sunlight and that accounts for the apparent excess output.

Resonating crystalline materials

John Hutchison of Vancouver, Canada was intrigued enough by a "self-reacting" material, barium titanate, to see if he could make a power cell that needs no batteries. He combined a special soup of filings of common metals and minerals, poured it into containers and applied a strong electromagnetic field while it hardened into a solid crystalline structure. He believes the crystal structures resonate in a way that taps into zero point energy. Proper wiring allows captured energy to flow out terminals, powering small clocks and calculators.

No moving parts

A mentor for Hutchison and other experimenters is Thomas Bearden, who points out flaws in classical electrical theory and suggests a new worldview. Bearden and associates patented a solid-state electromagnetic generator that uses a combination of magnets, coils and resonance processes. As you have learned, current moving back and forth through a wire near a magnet will produce a varying magnetic field which in turn can produce electrical current in another conductor. Bearden once said he needs to be able to hire a team of specialists to refine the device into a reliable generator of electricity.

A few of the inventors we know are finally connecting with business people and philanthropists who are promising to fund the inventions, but other inventors with very promising potential breakthroughs are still hungry for help. One company has a room temperature superconductivity project that could revolutionize energy generation and transmission. It needs development funding[239] as do other potential breakthroughs.

Russia is a leader among countries with scientists having made breakthroughs in revolutionary energy technologies. Certain inventors have developed high efficiency vortex generators of heat and electricity, operating on air and water, which involves steam bubbles collapsing into the energetic shock wave known as cavitation.

Novel waste-to-energy processes

Mining landfills and manure piles for energy sources (not only for methane gas) will not only lead toward increasing energy independence but also toward cleanup of the environment. A company in Connecticut has made it their mission to change the way the world views and employs discarded materials. StarTech's technology ionizes air to make an electrically conductive gas (plasma) and produce a lightning-like arc of electricity that transfers intense energy to wastes, including hazardous and lethal materials. The arc in the plasma plume within their vessel can be as high as 30,000 degrees Fahrenheit. Molecules in waste materials break apart resulting in a char that has other uses. The process is computer controlled and operates quietly at normal atmospheric pressure.[240]

A company in Pasco, Washington named Green Power Inc has developed a way to inexpensively convert biomass and household waste into high grade diesel fuel. The process can make fuel at about sixty cents a gallon. The product is a hydrocarbon, but on the positive side it consumes problematic plastics, waste oils, food and animal waste and wood chips. One ton of waste makes around 120 gallons of diesel.

Solar concentrators

Parabolic shaped dishes concentrate heat from the sun to an intensity of heat that can be used in generating electricity. The well known Stirling engine is one way to use that heat. It operates without internal combustion and uses a temperature differential to drive a piston and produce electricity. A company in Kennewick, Washington, Infinia Corporation for instance has a reliable version of the Stirling engine that operates on a range of heat sources and is commercializing a system using concentrated solar energy and their engine.

From the air car[241] to zero pollution vehicles, a downpour of technologies that can make a significant difference in our consumption of energy are raining onto Earth. The concluding part of *Breakthrough Power* begins with a chapter that considers a possible alternate path around the usual business route for moving a product from the workbench to the marketplace.

PART•IV

Who Hijacked the Energy Revolution?

Imagine a Comedy Hour in some distant galaxy. A space traveler has the microphone.

" . . . just back from Earth . . . Gotta love those humans! Wallowing in oil slicks and coal smoke and radioactive gunk. Brains big enough to invent a decent Hyperspace-Harvesting Power-Tap, but they wipe out anyone who does!

Why? Well, humans are told to think 'it'll collapse the world economy'! As if they're too helpless to figure out a different way to barter!"

If there were an extraterrestrial comic out there, we would indeed provide rich material. When humans are faced with a major change in worldview such as a new energy science paradigm, many of us stubbornly resist it although logic says the change would help humankind.

The unaware aspects of human nature contribute to creating world problems that are no laughing matter. Barriers to introducing "disruptive" energy systems to the marketplace have involved apathy, antagonism and political opposition more than technical challenges.

Keeping in mind the wisdom in the old saying "what you resist persists," the authors believe that the obstacles described in this section are best observed in a detached manner. That means noticing the reality of the obstacles without buying into anger or fear about them, much as a window shopper who is curious yet disciplined doesn't walk into the store and buy something that's not needed. You find out what's out there and then move on. And if you don't like what you see, you help create something quite different.

17

Misled Media

The ownership of broadcast and print media does touch on some of the core values Americans hold for freedom of speech, open and diverse viewpoints, (having) vibrant economic competition from a variety of sources, and local diversity.

—*Senator Maria Cantwell*[242]

There was nothing on there that remotely resembled what Watson did or what I did. It was a Rube Goldberg machine. . . . Never did they contact anybody. And they did it on national TV just to discredit the whole thing.

—*New energy inventor John Bedini*
regarding a device built for a debunkers' television program[243]

A nationally-televised satirical portrayal of John Bedini's innovative energy technology shocked people who know what he has actually accomplished. Unfortunately, millions of viewers likely believe the television show presented the objective truth, simply because it was depicted as such on TV.

John Bedini and experiment.

The myth builders

A strange segment on the television series *MythBusters* aired at a time when interest in the Bedini-designed "School Girl Motor" (see Chapter 19) was mushrooming. Bedini's Internet discussion group[244] had grown to nearly 500 people. Some were building the motor and starting to get results—charging batteries with the system and also getting mechanical work from their prototypes.

The portrayal pretended to take the viewer into a laboratory where Bedini's work was scientifically tested, but instead the "testing" was a mockery. The Massachusetts Institute of Technology (MIT) "expert", included on the show for his academic pedigree, must have known that the ridiculous televised device would never work. The way he—or whoever—built it resulted in more electricity being drained from its battery than was put back in for charging the battery. *Exactly the opposite effect of Bedini's real device.*

With exaggerated sarcasm, two frenetic characters on the show displayed what they rightly called a contraption. They presented it as having been built by themselves and the professor as a model of the Bedini motor. It wasn't. No one from the show contacted John Bedini to find out how to properly build a Bedini motor system, nor did they give any opportunity for rebuttal.

Instead of dealing with the motor being discussed on the Internet, the television characters made their version of a 1984 prototype, without even having the required magnets on the energizer part of the system. The thing shown on television had a flywheel, as had a model based on a Bedini circuit, built in 1984 by Jim Watson. Watson had publicly proven that it operated as claimed. Bedini cites additional omissions or changes such as coils wound incorrectly on the television contraption. Emphasizing words such as "fantasy" and "perpetual motion" the program's narration rushed through dialogue about a battery driving a motor that spins a flywheel[245] to charge wire coils that in turn recharge the battery. The dialogue ignored the fact that charging of a battery is a separated function in the Bedini motors. In actuality, Bedini's systems continually charge one battery or set of batteries while running off a smaller separate battery.

The show also gave the false impression that Bedini's claim for the motor is that it's an overunity machine—which the uninformed scriptwriters equated with perpetual motion.

In reality, John Bedini clearly tells people that the SG motor is *not* a free energy device. He carefully doesn't talk about overunity, but says it is a battery-charging system that demonstrates new principles. Bedini habitually distances himself from the "free energy" phrase for his own reasons.

Debunking shows are often backed by an irritating or exaggerated soundtrack designed to heighten the woo-woo impression, and the *MythBusters* episode was no exception.[246] The audience heard a loud mechanism winding down to a stop, and the effect of something fading and dying was reinforced by one of the characters muttering that it was slowing down just like his grandfather.

The sarcastic *MythBusters* script said the "Bedini machine" ran down—in a sudden burst of free energy. The dialogue gushed about overunity making you feel all warm and fuzzy like the Easter Bunny or Santa Claus does, replanting the idea that a machine which seemed to defy known physics would be a fabrication. The show's message was reinforced by an in-your-face close-up of one of the show's characters loudly spitting out the punch line that such a motor cannot work "because it's a fantasy!"

What it means to the rest of us

Why should you care whether an MIT professor cooperating with television personalities mangles the construction of a little motor system? We think you might care about being misled. After viewing that television show, the casual observer or an engineer just beginning to hear about Bedini's work would surely lose interest. With the power of a mass-audience television network behind them, television producers can appear to publicly discredit someone who in truth has world-changing inventions.

In contrast with that television segment, John Bedini is straightforward and candid. Co-author Manning has seen his battery-charging inventions working in his laboratory and enjoyed hours in discussions with him and his close associates. He is fully occupied with his business and family and building things that work, and has no time for debunkers' nonsense. He wants to help humankind, as did the inventor we meet next.

Betrayal and the hydrogen generator

Francisco Pacheco was a soft-spoken man with iron-hard resolve. He had developed a working invention that turned seawater into hydrogen fuel on demand, and devoted his life to getting it out to the public. Despite silence from government officials, oil companies, and utilities he persisted in writing letters to draw attention to his discovery. Mainly he worked on building and refining prototypes of his hydrogen-from-water generator. His experiences are archetypal of the obstacles hampering game-changing inventions. In this chapter we only describe his experience with big media. Small independent media treated him fairly, and journalist Karin Westdyk wrote in an environmental journal[247] a sensitive and thorough account of the inventor's life story.

As with other inventors, Francisco Pacheco believed that being the guest on a major television program would open the doors he needed to have opened. Television talk show host Geraldo Rivera did express interest after Pacheco demonstrated a power boat running on seawater, and Rivera wanted to do a show about Pacheco's generator. The idea was turned down by the television station involved.

After the Inventor's Club of America inducted Pacheco into their

Hall of Fame in 1980, the television news show 60 *Minutes* contacted him. The inventor was relieved and happy that his world-changing invention would finally be placed in front of a mass audience. Surely some viewer would help take the hydrogen invention to the marketplace, finally.

The television crew visited the Pachecos, and in a friend's barn the camera crew filmed the generator while it made hydrogen fuel for a Bunsen burner and for a torch that sliced through a three-quarter-inch steel plate. There were other visual demonstrations such as hydrogen inflating a balloon and running an electric motor.

One of the demonstrations involved a lawn mower, and that one did not go as well as it usually did. Pacheco had rushed out to buy a new lawn mower for the filming and didn't have time to test it out. As a result, the engine choked due to the excessive amount of hydrogen being produced. But this did not seem important in light of the success of the burner, the torch and the motor.

The 60 *Minutes* crew reassured Pacheco that they had enough material to present an entire show with the successful demonstrations.

Karin Westdyk wrote, "Later when the show was aired, Pacheco was devastated, as the show had a completely different focus. The only demonstration aired was the lawn mower, and it was used to provide an example of an independent inventor's non-working invention."

Although advised to sue the program for misrepresenting his work, Pacheco discovered that the legal cost would be more than he could afford. He had already mortgaged his home to finance prototypes and demonstrations.

What would motivate a major television network and high-profile "uncover-the truth" news program to do what amounts to a "hit job" on a brilliant, sincere inventor and his remarkable technology? Whose interests does it serve to smear the credibility of a gifted innovator and his remarkable clean-energy discovery?

Prejudices

We have not sat in any smoke-filled rooms where decisions are made that result in manipulations of news about energy breakthroughs. However with friends in local media and with Manning's past experience working in newsrooms, we see that certain attitudes have filtered down to the level of reporters. Even though the following example has to do with the topic of so-called subtle energies, the attitudes are universally applied to alternatives in general.

One of our friends, a popular reporter for a daily newspaper, does not realize she is biased. It came to our attention when she told us that her editor had assigned her to write about a local group involved in something called "vibrational healing" (similar to the "therapeutic touch" method practiced by many nurses on their patients). "I'm trying to get the goods on them," she told Jeane heatedly. "They'll slip up soon enough, then I can expose them for what they are."

And what would that be? She didn't know. At that point she had not met the people involved and admittedly had not done any reading on the topic either. Regardless, she was certain that the healing practices were a scam and the practitioners must be misguided or fraudulent. Why? Because "the experts" who speak for the established medical community at best ignore such non-conventional practices, or when questioned will utter an amused dismissal.

The newsroom culture blinds our friend from seeing herself as prejudiced. In her mind she's a fearless writer of exposes, tackling the scammers who prey on the uneducated public. That's a worthy mission; vultures should indeed be exposed for what they are. Trouble is, her advisors—and her editor—seem to have a reflex attitude that automatically scorns people who report revolutionary advances in health or in energy inventions if these breakthroughs don't come from within the current academic worldview.

Defense of media

In defense of journalists, we have found that a minority of inventors of energy devices are off-the-scale eccentrics, self-deluded "messiahs", or outright hucksters, and an extended encounter with one of them can be annoying, tiring or even frightening. It is not surprising that a few lone inventors become unbalanced by thinking that they alone can save the world. Unfortunately, that small percentage of the energy researchers contains the people most likely to badger an editor or reporter. Sometimes the automatic scorn reaction has protective value for journalists themselves by limiting the risk to their personal and professional credibility. New energy advocates only ask the media to objectively look beyond the personalities and much deeper into the field of quantum-leap breakthroughs.

Uneven treatment

Wingate Lambertson, PhD is on the respectable side of the scale of inventor personalities—well-dressed, well-spoken and far from eccentric. However he is frustrated with the media. Lambertson is a Florida-based physicist and inventor of a method to tap into what is called zero-point energy or vacuum energy. He describes his reaction to a small example on a television program about astronomy. A famous physicist explained that mysterious "dark energy" exists throughout the universe and then quickly added that it "cannot be collected."

"As an individual who has witnessed the collection of this energy, which is known in the field as zero-point energy, I was appalled," Lambertson commented. "Why did the authors of that program find it necessary to add that disclaimer?[248]

The disclaimer that the energy cannot be collected was peculiar in that the statement had no relevance to the topic of the program, which was astronomy. What compelled the program producer's need to preempt any viewer's potential questioning, "Can we harness some of that universal dark energy?"

Paid debunkers?

In Ottawa, the founder of the Planetary Association for Clean Energy, a non-governmental organization (NGO) listed with the United Nations, shares what he has seen. "Some individuals are actually *paid* to be debunkers and they belong to 'institutes' that are actually spin-doctor mills. If you were to see their offices, you would be amazed at the number of outfits that are listed next to the doorknob. They *appear* to belong to respectable outfits and we infer by association that they actually belong to something else than what they actually do."

". . . How much they are against something depends on the size of the retainer fee. So do not take them seriously. Unfortunately, some of the media does."[249]

Genial skepticism

The authors' colleagues Steve Kaplan and Brian O'Leary crafted a well-referenced article about new energy research, aimed at a mainstream audience. Kaplan called on a former classmate from graduate school at Cornell University who at the time of the call was executive editor of a well-known mass circulation magazine.

"We brought our article to him; it sat for the longest time. Finally we get this strange letter. '. . . well-written article, exciting, but you're too much of an advocate. We really wanted something that would express genial skepticism.'"

The game rules are not evenly enforced; the magazine has carried many advocate-type articles, Kaplan noted. "I suspect that what happened is it got to their science editor and the science editor was tied into the scientific establishment that says 'it can't be.'"

Television un-documentary

One well-connected independent writer for television who knows which producers to pitch ran into a wall when she proposed a documentary on breakthrough energy inventions. Only one producer wanted to take on the project, but on the usual terms. "Let's make it fun and innocuous, with a host who's a fun debunker."

The ubiquitous framing of stories about outside-the-box science filters down to the local newspaper level. That usually becomes evident when a town hosts a conference on such topics.

Byline fever

Colorado Springs, Colorado was the site of an "Extraordinary Science Conference" in the 1990s. We have a copy of the local Gazette Telegraph newspaper's coverage titled "Totally Quirky Science" in which reporter Rich Tosches leads off by saying that just a mention of the annual conference conjures up an image. "Men flying here from parts unknown, their hair standing on end from one encounter too many with an amp surge, the cuffs of their pants and eyebrows smoldering from a voltage regulator experiment gone horribly wrong."

Tosches apparently looked for the wildest attendee at the conference, because he gave a good portion of article to a man who "roared into town in his Ford station wagon that has a leak in the transmission and a mattress in the back." Further details showed the man was indeed eccentric enough to warrant columns of newspaper space.

Following the unwritten rules for dealing with unconventional energy research, the reporter interviewed a skeptic. A University of Colorado physics professor begged that his name be withheld, adding more humor to the reporter's piece. ". . . these Tesla people, well, I have seen some of their work and very little of it looks like good science. I believe they are more of a cult than a legitimate scientific group. They don't get mentioned very often in conversations among my colleagues . . ."

Tosches had a few more laughs to share. Nikola Tesla was by all accounts a genius. "And at times a loon." Tosches next pounced on the most outrageous of the pamphlets on tables at the show.[250]

In short, the reporter wrote a by-lined article which probably provided amusement in the newsroom and a prized sample of humorous writing in his portfolio. Meanwhile scientists and in turn students at nearby universities have been persuaded that there is nothing of value at a Tesla conference and vow to stay far away.

Waste of innovators' time

In that same decade a number of highly credentialed scientists were among the line-up of speakers at an International Association for New Science public meeting in nearby Denver, Colorado. Astrophysicist Brian O'Leary, PhD was among those who were interviewed by a writer for a mainstream East Coast newspaper. He and his colleagues took the time to explain in layperson terms some basics of new science and quantum-leap energy innovations. When the writer's piece was published, however, there was no mention of those concepts. It had turned into another light and dismissive piece about Nikola Tesla.

Gate-keepers of science

Physicist and engineer Eugene F. Mallove, PhD, often spoke out about scientific censorship. Until he was murdered in 2004, he wrote a magazine column titled "*The New Heretic*". With the permission of *Atlantis Rising* magazine we quote extensively from one of his columns that dealt with censorship.

> . . . how fantastically better off civilization would be, were we allowed to use collectively all our faculties and power of reason.
>
> Isn't science supposed to be one of the most liberating endeavors? How can I claim that we are not being 'allowed' to use all of our faculties and powers toward making a better world? Easy! If there is even one choke point at which appropriate information about scientific discoveries are withheld or diminished, the community of scientists and the supportive citizenry who fund

their work publicly and privately are defrauded. Sadly, today such a chokepoint exists: it is the routine censoring of scientific information that does not conform to the dominant scientific paradigm of the day.

Yes, it is the Internet age and all kinds of heretical scientific information exists and can flow freely in that ethereal world of rapidly moving digital information. On the Internet one can find large stores of information about cold fusion/LENR (low energy nuclear reactions), hydrino physics, aether (vacuum) energy, heretical astronomy and cosmology, complementary medicine, scientific evidence for 'paranormal' phenomena. . . . a cornucopia of scientific heresy, albeit of uncertain quality. And that is the crux of the problem: how is the average citizen, whether scientifically trained or not, to distinguish good from bad scientific information?

Because we are in a transition stage in which the credibility of various sources of information is still being sorted out, we are stuck for now with a system in which certain influential scientific publications are deemed to be reliable authorities on the status of scientific paradigms.

For example, *Science* and *Nature* magazines have become over the past half century dominant influences on what is to be regarded as 'acceptable' and what is not. It is not surprising that the powerful science journalism industry has grown to regard such publications as nearly the final arbiters of truth—though at the same time that science journalism community acknowledges instances of scientific fraud, later discovered, that managed to get published.

So the influence of such publications is far, far greater than their immediate circulation numbers would indicate. If *Science* and *Nature* or *Physical Review*—have declared certain topics to be off-limits and questionable, you will rarely or never find articles about those topics in the *New York Times, Scientific American* or other organs of more general audience science journalism.

But because the general population has assimilated the idea that articles submitted to the archival journals are fairly treated in 'peer review', the 'absence of articles' on a particular controversial phenomenon is taken as evidence of lack of evidence for the phenomenon itself.

Yet there exists controvertible proof on a broad range of topics that proves this belief false. These are the very specific instances in which the overt censoring of science was practiced by mainstream scientific publications.

Richard Milton in his book *Alternative Science: Challenging the Myths of the Scientific Establishment* says we are in an era in which a scientific fundamentalism has infected many universities and the cold fusion controversy was a prime example. Throughout history there have been times in which academic intolerance hardened against new discoveries and ridiculed and rejected them for unscientific reasons. However, he expects the self-regulating nature of science will balance it eventually because "an ounce of open-minded experiment is worth any amount of authoritative opinion by self-styled scientific rationalists".[251]

How to start the education

Steve Kaplan once gave pointers on how to "start the education about the gentle transition to a new energy economy." With humor he suggested a "water torture brigade" for dealing with the press: "Get informed. Keep knocking on their doors. If the environmental editors don't fulfill their promise. . . . then go to the editorial board. If that were done to major newspapers throughout the country by well-informed persons, there would finally be coverage."

Exit the mainstream

As people abandon the traditional media[252] in favor of "online" their choices for news sources are multiplying. Others realize that mainstream media still influence voters and other decision makers in our society, so citizen groups are attempting to change some practices of the consolidated media empires. The Media Action Grassroots Network is a coalition of regional organizations working for media justice.[253]

Seattle-based Reclaim the Media for instance works to change media policy to "favor the public interest, rather than a powerful elite". The nonprofit organization teaches media literacy—education in "how news can be shaped by journalistic habits and by powerful commercial and political interests."[254]

As far as we know none of these groups have championed new energy science. We believe these gaps between worldviews held by citizens will soon narrow because it is in citizens' interests to seek new-science literacy and news. The next chapter underscores this view, revealing and documenting the gap between the environmental community and new-energy researchers, as well as depicting other challenges faced by the energy revolutionaries.

18

Hitting the Wall

Most of the really new inventions come not from academic teams or corporations but from the lone, independent, fiercely creative people.

—TOM BEARDEN, *physicist and author*[255]

Astrophysicist Eric Lerner[256] had an unhelpful surprise shortly after speaking at a conference on plasma science.[257] Expecting that his team would be congratulated on their discovery, he announced their recent findings. The experimental results indicated that a compact non-polluting fusion device called plasma focus had opened the door to a new energy source that could theoretically be a hundred times cheaper than oil and gas. The experiments were done in 2002 at Texas A&M University and the project was funded by NASA's Jet Propulsion Laboratory.

Other scientists did praise the work of Lerner's research team, but a Los Alamos National Laboratory manager threatened to fire two members of the team if they didn't reject the findings and say the results were false. Funding for the project was cut off, and the media ignored the discovery.

One physicist on the team had not even been working for Los Alamos during the research, yet his new employer at the national laboratory demanded that the physicist distance himself from the report's comparisons regarding the tokamak, a donut shaped hot fusion device that confines and heats a gaseous plasma by using electric current and powerful magnetic fields. The report from Lerner's team showed that their plasma focus using hydrogen and boron as a fuel would be far superior in important aspects.

Hot fusion no threat to oil

That was apparently not what some people wanted to hear. For more than twenty-five years the tokamak design has been the large and expensive centerpiece of the American publicly-funded hot fusion effort. And conventional hot fusion is no threat to the oil industry.

A member of Lerner's team who works at Texas A&M University was also pressured to advocate removal of the comparisons to the tokamak. Lerner was outraged. Forcing scientists to recant or be fired interferes with honest sharing of information. He was angry enough to take his case to the public and seek donations to continue the work. The money needed to take the plasma focus to the next stage was $500,000—a miniscule amount compared the millions given to hot fusion over the years.

Lerner told readers of the *Progressive Engineer* that plasma focus does threaten existing fossil fuel multinationals because focus fusion reactors would be able to produce energy at a price equivalent to oil at *a dime a barrel*. A scenario of a focus fusion reactor for each town, each producing about twenty megawatts, would inhibit corporate control of energy supplies.

Where's the red carpet for breakthroughs?

Waiting for 'others'

The piston engine has served humankind for two centuries and automotive engineers are well aware of its environmental, economic and other limitations. Now a breakthrough made in Quebec begs it to retire.

While on a sailboat with his wife and children during a three-year sabbatical, thermonuclear physicist Gilles Saint-Hilaire PhD designed what many technical commentators believe should be the next generation of rotary engines.[258] It's an efficient, compact machine with no crankshaft and with benefits to economies and the environment.[259] Working with a process called photo-detonation[260] he has designed further generations of beneficial replacements for conventional engines.

Ready to present the good news, he and his team took the technology to major auto shows.[261] Reactions puzzled Saint-Hilaire.[262] Individuals did express curiosity about the Quasiturbine but representatives of companies didn't seem to bother to look at it. A typical comment was "We've seen so many fly-by-night engine inventors . . ."

But that excuse was not the full story. The advice given to his associates behind closed doors was blunt: a new technology should be introduced to the car industry through the credible network of universities and engine consultants, not by the "little guy himself." The inventor wryly concludes that the value of a technology depends on who owns it.

Some corporate representatives expressed willingness to look at the new engine if they were not the first to step forward. They said, in effect, "If you interest others, we will show interest."

Universities, national laboratories and government departments didn't even want a free demonstration of the breakthrough engine. Saint-Hilaire speculates about the reason. "The advantage of ignorance is being able to claim, 'We did not know, we are not to blame!'"

If behind-the-scene pressure from vested interests deters anyone from investigating breakthroughs, vested interests appear to be winners. But only in the short term, Saint-Hilaire adds. In his view the long-term losers are smaller countries including his own that are unwilling to "affirm their intellectual capability" against economic powerhouse nations. Prosperity will go somewhere else.

Canada mobilizes people to save factories in danger of closing, what about mobilizing for a beneficial technological revolution? Saint-Hilaire asked. Despite the best efforts of Quasiturbine supporters,[263] "no government official including (those in) energy and environment has ever suggested to the car industry or to a university or national laboratory to look at the QT." In protecting existing industries without at the same time promoting a next-generation industry, Saint-Hilaire said, decision-makers are like farmers who pay attention only to the crop, not to the seeds.

Saint-Hilaire puts into plain words the evasive attitudes he has encountered. "We are happy with technologies we already have; we don't need any engine revolution."

The authors empathize with his disillusionment, but hope that soon he and other leading innovators will have significant funding to advance their inventions into the marketplace. We see signs that justify hope, both in the public's awareness of the need for change and in financiers' willingness to support breakthroughs. But the quantum leap energy technological revolution had not really begun yet. Vested interests can't be totally blamed for that.

Preserving 'expert' status

A revealing incident happened when new energy experimenter/historian Peter Lindemann was working daily with inventor John Bedini. It wasn't unusual for engineers to invite themselves to Bedini's laboratory. This time, engineering experts employed by a large corporation convinced the inventor to allow a visit. Two of them traveled from Germany and one from California to meet the inventor at his workplace in northern Idaho. The experts unpacked so much testing equipment that it took several hours for them to set it up.

Bedini had made clear that he was not claiming more output than input to his rotary magnetic system. What the visitors saw was jaw-dropping anyway. The little magneto wheel was sending charge to a battery only one-tenth of the time, but the battery voltage was rising slowly, but continuously. The engineers' instruments showed that the battery continued to charge itself during the intervals in-between the charging pulses. The system charged one of two batteries faster than the second one discharged while being used to run the motor.

Once they realized what was happening, they just put down their meters and stopped the tests. Then, they stepped outdoors to huddle as if to discuss what to do. When they walked back into the workshop they announced "We have a problem. If we go back to the corporation and report that it works but we don't understand *how* it works, we may be fired because we're supposed to be the experts and we are supposed to know everything."

"On the other hand if we report to our bosses that it doesn't work, then we can just go back to our jobs and nothing will change for us."

At least these gentlemen were honest with themselves and with the inventor. Other inventors say they have observed similar inner conflict in experts who go from saying "that's impossible" to silence and then to "I didn't see anything."

Many inventors of potential breakthroughs in the energy field have told us what a shock they experience when they first encounter disinterest from business, government or universities. Needing help in testing, developing and marketing the new energy generators, the innovators often become disillusioned by the indifference encountered.

One innovator described a "chicken and egg" problem. Inventors without access to laboratories can't produce demonstration devices without the money to build them. And they can't seem to get the funding until they at least have that device "firing on all fours," As a result many brilliant concepts languish.

Even PhD scientists find themselves in such situations. The needs for funding and for patents and for having published their research published in scientific journals are intertwined. "We can't take you seriously until your work is peer-reviewed and published . . . . We can't publish your paper until your work is endorsed by an accredited university."

The science maverick without a university degree has few opportunities in such an environment. John Hutchison recalls the time when a businessman who was interested in funding Hutchison's research took the inventor and his tabletop "crystal converter" to meet officials whom the businessman believed would help. "We went downtown to the Canadian government research and development people, to show them the big converter. We plunked it down on their office table. There was just apathy. They commented 'Well that's interesting . . . . You might find some luck with it somewhere, but we can't do anything.'"[264]

Greens shut their door too

Eventually innovators turn elsewhere. "If the government/university/

corporation doesn't want to help me, I'll go to the 'Greens'. The enviros want to clean up the atmosphere; we can be allies. They're against oil spills, coal smoke, polluting cars. We have good news for them!"

Common reactions from the environmental community, however, are yet another shock to the innovator. Again, backs are turned and phone messages unanswered. And when a response does occur, it is often some version of "technology ruined our planet; free energy will just hasten our demise. Buy a used bicycle and turn down the thermostat."

When the authors hear those stories, often our sympathies are with both the inventors and the environmentalists. Both sectors contain idealists who care deeply about the web of life on Earth and who could be collaborating instead of seeing "the other" as antagonistic. Perhaps considering where a differing perspective came from could help bridge the gap. It's more than a small gap; a virtual chasm opens up between neighbors when one is a serious "deep ecology" environmentalist and the other neighbor joyfully praises unconventional energy generators as a source of energy *abundance*.

Where did these differences in worldview originate? As we saw in Chapter 2, a brief look at history reveals one root cause of differing attitudes toward the concept of abundance. Abundance is commonly regarded as either the effect of greedy self-interest or an impossible dream for the consumer.

Encounters with mindsets

The Canadian government's environment ministry went on the road from city to city one year with a presentation about the problems of emerging climate change, air pollution and what we can do to halt such problems. Its to-do list for the public was standard tighten-your-belt-and-conserve so that polluting energy is used less. Bicycle, walk, recycle. Way to go, but there's more that we can do, co-author Manning thought while attending a presentation. Traditional energy alternatives were mentioned in passing. Manning went up to one of the scientists at break time and told him about some of the new energy developments that are totally non-polluting.

"I'm against technological solutions!" he said.

She walked away from that closed mind. He apparently wasn't willing to hear that the problem isn't technology per se; the problem is the type of technology we use.

A Greenpeace researcher in eastern Canada bristled at Manning's suggestion that some government research funding could go toward developing the new energy inventions. He protested that any dollars diverted to researching undeveloped science would take dollars away from proven alternatives such as solar power. "The scarcity paradigm rules," she thought wryly.

After reading Freie Energie, the German edition of The Coming Energy Revolution, a trio of Greenpeace members in the Netherlands tried to get their organization's leaders to look at new energy inventions in Europe. They reportedly failed to rouse the leaders' attention because "inventions" and "technology" are turn-off words.

Groups such as Earthsave rightfully fret about consumerism because it does deplete scarce resources. Cheaply made products fall apart and become cast-off garbage piled onto Earth.

However, things could be engineered to last much longer if designers didn't have to worry about embedded energy (the cost of energy used to manufacture the product) and more materials could be cost-effective recycled with cheap energy. Today it takes more energy to recycle some materials than the power that would be saved by not making things out of virgin materials. But societies will be able to recycle more of their used metals, plastics and glass when recycling depots use fuel-less energy sources. Technologies also exist for turning waste into building products and getting electricity at the same time.

'Don't give up'

Humankind has a choice between the old way—of oil wars, smog, coal dust and nuclear waste—or a new way. Decision-makers can choose to gradually embrace the revolutionary abundance of new power sources and use the wealth of energy to clean up the planet and heal our economies. We can still leave a light footprint on the earth, if enough people decide to support that new scenario and do it right.

It's easy to dismiss such talk as being a dreamers' manifesto, while governments spend lavishly on subsidies for destructive energy choices, wars to control fuel supplies and on secret projects. The classified budget of the United States Department of Defense nearly doubled during the George W. Bush presidential administration. In 2008, $32 billion went into the secret world of unacknowledged advanced science and technology.

Meanwhile progress on Graham Gunderson's project for instance has been much slower than it would have been if he could have hired a team of helpers, or even one assistant. If the project were well funded a team of specialists could speed up the research and development phase. In a number of countries, brilliantly creative individuals and privately funded teams struggle to fund research and development of their quantum-leap clean energy breakthroughs. As stated by Eric Lerner, independent engineers and scientists whose work could benefit the rest of the human family frequently found that their efforts were hampered by corporate interests and the governmental policies designed to serve those interests.

Moray B. King, electrical engineer and systems engineer you met in chapter 6, has been around frontier physics for thirty years. He lists eight ways that revolutionary energy inventions have been blocked before they had started on the road to the marketplace.[265]

- academic suppression
- blocking of funding
- blocking of patents
- litigation
- threats to the inventor
- property destruction
- framing the inventor with a crime
- assassination

King says the problems come from three main communities—academic, business and 'black operations'. The academics' tactics are usually benign—ignoring, ridicule, peer reviewers' rejection of papers, shunning and accusations of fraud.

Industrialists and business people have less of a problem facing the fact that a new paradigm is coming from outsiders. However they don't want it to happen if it means they would no longer have a monopoly or would have to retool their factories.

The third source of suppression is the secretive world of projects that are called "black" because even presidents of the country cannot open their records or even know they exist. The military/industrial complex receives billions of dollars for "unacknowledged special access projects" but doesn't account for how those taxpayers' dollars are spent. King points out that some insiders in the black operations world have honorable motives; they sincerely believe that national security is served by not allowing breakthrough new energy discoveries to be developed for the public's use.

In light of these realities facile cynicism can erupt contagiously, but cynicism is a variation of giving up. Fatalistic acceptance of an unacceptable situation is *not* the only choice humankind has. Instead people have numerous empowering choices. The next and final part of this book looks at some ways that we, the people, can help bring out the new energy breakthroughs that have the potential to transform Earth.

PART • V

You Can Help Reclaim Choice

Do we want a new-energy future of breakthroughs being widely used for the good of all life? If you ask most people, they would agree that they do. Are citizens willing to demand that new energy future? That is the important question. If we do not demand a change for the better, then a global switchover to a world economy based on clean energy will remain on the far horizon, a vaguely seen dream.

Resistance to change is in proportion to the size of the entity that must learn to deal with such change. In this case, the entity is colossal—a world economy based on fossil fuels. It is a formidable entity, with an outlook weighted toward military-industrial alliances. Yet the people steering that economy *could* find the wisdom and courage to change priorities from maximizing financial profits to cooperating to save their world. Already some decision makers within the power structure are recognizing that their own children's survival is threatened by decisions made in the twentieth century.

We the people don't need to wait for vested interests to wake up before taking action. This last section of *Breakthrough Power* gives some suggestions and in conclusion aims to set the tone for a grassroots movement into a new energy civilization.

19

Opening Doors

There isn't enough uranium to replace fossil fuels.

—*Richard Smalley, Nobel Prize winner*[266]

Meters only measure wasted energy . . . There are no meters to measure this radiant current. When you catch it, it has the power of the universe and beyond.

—*John Bedini, Electronics technician*

Mr. (Bob) Boyce is a true humanitarian and has placed his system in the public domain in the hopes of helping the planet reverse global warming, end oil wars, and deliver free energy to the poor peoples of the world.

—*Michael Couch, writer for Pure Energy Systems News*

Shawnee Baughman was nearly ten years old [267] and she wanted a science fair project. Not the boring motor shown in the library book, made out of corks and match boxes. "I'll get you the parts for a better motor," her father promised.

Fortunately for Shawnee—and the world—her father's business was located near that of electronics innovator and new-energy pioneer inventor John Bedini, who agreed to instruct Shawnee after school until she had built a motor. A real motor. With him alternately explaining over the phone or demonstrating, and Shawnee doing the work, an innovative motor and battery-charging device was completed in time for the science fair. The judges gave her top prizes, but what pleased Shawnee most was that the other kids liked the motor and voted it Best of Show.[268]

Shawnee Baughman and her prizewinning school science fair project.

Hope for the future

Bedini later told Jeane Manning that he helped Shawnee because children like her represent hope for the future, and the demonstration "drove science teachers nuts—to see a little motor made of plastic with no return paths for the magnetics." (Bedini's website[269] and Thomas Bearden[270] explain technical details beyond the scope of this book. However, it does fit into the scope to tell you that Bedini's lifelong approach of learning from nature is paying dividends. His unusual battery-charger products on an industrial sale are being sold commercially,[271] and they involve the "radiant energy" discovered by Tesla and T.H. Moray. Bedini says that radiant energy surrounds everything we do; plants and trees use it constantly as a life force energy.)

Bedini's actions—taking the time to help a child—ended up mobilizing researchers around the world, as with the hypothetical effect mentioned earlier in which a butterfly flapping its wings on part of the planet can create an atmospheric storm elsewhere. The challenge went online, "If a school girl can build this, why can't you guys?" As a result, discussion groups on how to build a "SG Motor" mushroomed (see chapter 17).

Bedini is doing his part to take energy breakthroughs from the laboratory to the people, despite challenges that were placed in his path. There is more that he could do if the political environment were welcoming a real energy revolution. However, the authors see grassroots support as more readily available now than in 1985, the year when Bedini first stepped onto the new energy stage in public.

In 1985 radio host and new energy supporter Bill Jenkins of California used his guest speaker's spot at a Town Hall Forum in Los Angeles to announce a "free energy" device made by then-37 year old Bedini and his colleague Steven Werth. The two demonstrated a small generator with 180 per cent efficiency.[272]

The audience included public utility representatives and investment brokers. Bedini told the forum that he planned to make the generator available to the public at a low cost, instead of selling to the highest bidder.

Civic officials at that luncheon meeting did *not* light up, so to speak, as they sat at the table with a free energy demonstration including light bulbs strung past their plates. Bedini recalled growled demands to remove the damn bulbs so they could eat.

Beyond threats

A few weeks after the public demonstration two thugs showed up at Bedini's shop, obviously hostile toward his efforts to unhook from the power structure. They had the muscular appearance of bodybuilders, and pushed him up against the wall. One gestured toward the street. "Is that your vehicle out there? . . . You're gonna keep on buying gas for it, aren't you?" Bedini decided not to risk his family's safety; his paradigm-changing energy experiments dropped out of public view.

Now almost twenty-five years later, the authors wonder if those thugs and their employers are feeling the heat of climate change and considering a change in tactics. Another inventor who has received unwelcome visits, incidents which left him with a broken tooth one time and an injured spleen another time, called Jeane Manning recently. He said that a recent telephone call had informed him that they were now going to allow the "free energy" innovators to begin bringing their inventions out to the public—within limits that he had been warned not to disclose. Who are 'they'? she asked. He replied that bankers are a driving force among an international group of behind the scenes controllers.

Creating a new culture

Meanwhile, new energy researcher Terry Sisson[273] of Colorado is one of the many people, like Bedini, whose seemingly small deeds locally have an effect that ripples around the world. A few years ago Sisson was reminiscing with his sons' friends about the hands-on construction and other projects he had worked on with his sons and how the boys' many friends had often joined in. These boys, now grown, told him how much that involvement with practical skills had meant in their lives. "Maybe other youth would like to learn skills such as building and designing products," Sisson said.

He approached a local school where the idea was warmly welcomed. The next stop was to solicit donations of materials from manufacturers and volunteers to help the youths make parts in their shops. They all agreed to help. "MadeByKids" quickly grew to include an after-school, experiential-learning program. Sisson and Phil Groth, a humanitarian entrepreneur from Australia, became co-founders of the project. When Sisson traveled to Russia on a lecture tour about Sustainable Communities, he approached schools to test out the MadeByKids concept. School headmasters jumped at the chance to have their students collaborate on projects with other youth around the world. A few years later Groth and Sisson committed to making the MadeByKids Foundation their life's focus.

Sisson is a member of the board of directors of the New Energy Movement. His vision of youths making friends with fellow project builders around the world is manifesting and is helping birth a new planetary culture which encourages cooperation instead of competition, understanding instead of alienation. Those are qualities of a civilization in which quantum leap inventions can be welcomed and flourish.

Even before that envisioned new culture becomes reality, some of the new energy devices and systems will percolate out of inventors' workshops, including those systems molded at least partially by the hands of youth.

How to bring out an energy breakthrough

A wide gulf exists between inventing and having a product for sale. Being a brilliant inventor does not necessarily include being an astute business person capable of attracting capital investment for developing and manufacturing a product. Therefore the topic of how to get to the final-product stage comes up regularly among innovators.

Many venture capitalists today are eager to invest in the "green tech" sector, so you would think that inventors of energy breakthroughs would already have the money they need for an engineering team, a well-equipped workshop or laboratory and a project manager or whatever is required.

The reality is that experimenters who passionately explore previously unknown principles typically exhaust their own savings and are loaded down by personal loans long before their inventions approach the pre-manufacturing stage.

Where are the venture capitalists in this picture? Most of them won't fund an "outside the box" technology unless it is already fully developed. They naturally want the "low-hanging fruit," a product ready for picking up and taking to market easily. Concepts that explore new physics principles or otherwise lack the blessing of mainstream science are considered too risky to fit the profit objectives of the investors. This fact is one reason why the *conventional* renewable energy technologies of windpower, solar, and biofuels are attracting lively venture capital attention lately, while unconventional breakthrough technologies languish. Old and familiar feels comfortable and less risky.

This is a major reason why there is a need for visionary philanthropists whose motivations go far beyond profits. Humanitarian funding of new energy could swifly fill the sails of many deserving inventors and propel their breakthroughs out of the backwaters and into the mainstream.

The standard business model has been the only approach to the marketplace that most people have been aware of, including new-energy inventors. It *has not worked* for new energy technology. At the time of this writing no one can drive to the local big-box store and buy a breakthrough energy technology. None of the quantum-leap inventions are even close to the point of being available for sale at a normal consumer outlet.

Open sourcing—a new era

John Wong is a retired engineer who senses something encouraging happening even in today's troubled era. In the electronics and software scene he witnessed a change in how processes develop in today's culture. "The days of serial processes coming from the top down are over. We now live in a parallel Open Source culture with its Web 2.0, Linux, Wikipedia, YouTube. This movement is still in its infancy but growing fast in the Internet world."

The open source method Wong mentions, combined with the power of the Internet, may be the most effective way to bring a quantum-leap energy system out to the public. The concept of *open source* first flourished among computer programmers who made the source code of certain computer programs available to users or other developers, who could then use or modify the code as they saw fit. When those users made improvements, the improvements were built into the next versions of the programs.

Among inventors, open sourcing means that no patents are taken out on the technology, because the inventor is not trying to protect intellectual property. Instead, he or she is sharing it and wants to publish the information widely to encourage its adoption and on-going development.

For those inventors who, out of concern for potential loss of future profits, simply cannot accept the idea of "giving away my invention", an alternative hybrid approach to open sourcing is possible. The inventor may decide to patent and then grant a free license to the public, a license that permits construction and operation of the device for personal use only. Businesses with a commercial interest in the technology would still pay a licensing fee.

Standard model, business as usual

Let's compare open sourcing with the standard business model whose sequence goes something like this:

The inventor discovers a new approach, develops a proof-of-principle experiment or an early prototype, applies for a patent to protect his or her intellectual property, and connects with "money people." Those potential funders view a demonstration and review the inventor's business plan then agree to finance the development of the invention to the commercial prototype stage—if the inventor is lucky.

However, that investment capital is not free-of-charge. In exchange for financing the technology development, the funder gets a share of the company that arises from the invention and part of future profits. In some instances, the funder may receive a major or even controlling interest in the venture in exchange for bringing professional management to the company. Once the invention has been advanced to the status of a viable commercial product, the company may do its own manufacturing, sales, and marketing or arrange licensing and royalty agreements with other companies who do those tasks. They all hope the product makes an impact on the market.

By this stage, often the inventor and funder want to sell their company or merge with an established larger one. If this "exit strategy" is successful, the funder makes a substantial profit on the investment. The inventor benefits from sale of his or her own stock, ongoing royalties, and perhaps employment in the acquiring firm in some technical or management job.

Under the normal business model, the technology is manufactured in a centralized location or locations and the product goes out the door and arrives at consumer outlets through standard marketing and distribution channels.

For a breakthrough energy technology, this business model is vulnerable to obstructions at several different gates. Whether it happens in the patent phase, the funding phase, the business team phase, the merger and acquisition phase or in legislation and regulation, there are a variety of ways for antagonists to sabotage efforts to market a new disruptive-of-the-status-quo technology. The status quo in this case means the fossil fuel and nuclear industries, and related mining and infrastructure industries that together bring trillions of dollars annually into the pockets of their shareholders.

Resistance to change in any sector is in proportion to the amount of wealth and political influence held by that sector. Consider the stakeholders who profit financially from the energy industries:

- private individual and conglomerate corporate owners and leaseholders of extractable natural resource wealth

- owners and builders of power plants, refineries, pipelines, transmission lines, service stations, and internal combustion engines

- millions of miners, oilfield and gasfield workers, refinery and power plant operators and engineers, gas station attendants, and thousands of related contracting companies

- governments that receive tax revenue from fuel and electrical utilities.

The tentacles of the energy giants touch nearly every economic level and nearly every community on the planet. Vested interests abound. It's no wonder that the lives of soldiers are sacrificed to protect a resource that resides under someone else's country.

Obstructing a newcomer who is seen as a competitive threat happens all the time in the business world around standard technology. When those well-developed interference tactics meet up with something as highly leveraged as breakthrough energy, the most world-changing technology that's ever been launched on this planet, the obstruction gets serious. Vested interests come out like a pack of wolves, instead of individual jackals here and there. Inventors have been have been denied patents, slapped with government gag orders, had their laboratories vandalized or in one case burned and have been entangled with business partners that ruined their opportunities and then left the scene. Some inventors have been threatened or beat up.

Critics of the open source concept insist that the standard business route is the only thing that will work. However, these critics generally have no track record for working with a quantum leap energy technology. There is no precedent for successfully launching a breakthrough new energy technology using the standard business model since the time of Edison and Tesla. (Nuclear power generation was ushered in under the close auspices of the military and other government agencies.) The decades throughout which that standard business model was tried were a history of obstructions that prevented the launch of energy breakthroughs into the public domain.

Open-sourcing breakthroughs

The concept called open-sourcing could help people finally break out some of these energy technologies into the public domain. It is possible to do an end run around the obstructionists or circumvent the battle this way. The following is what could happen: The first part of the business model —invention and prototyping—stays the same.

However the next part, the approach toward patenting, is radically different with open sourcing. Instead of being secretive to protect his or her discovery during the lengthy and expensive patent process, the inventor widely publicizes the detailed construction and operating information on the Internet for the world to use and perfect. Spreading the plans across the planet de-targets the inventor. What's the point of trying to obstruct or suppress someone once their secrets are out in the open?

How to organize an open-source launch

A coordinated launch of the technology could have highest impact by a combination of press conferences and Web broadcast. Press conferences at which the invention is demonstrated for the media could be timed for simultaneous broadcasting so that Los Angeles, Toronto, London, Tokyo, Sydney and other key cities get the news at the same time. Meanwhile an Internet launch would send the how-to-build plans to media outlets, academic institutions, social networking sites, civic organizations, environmental groups, science foundations, technology chat rooms, and religious organizations. All would receive detailed information on how to replicate the device. Information would literally go to millions of people at light speed. The people of Earth would have it in their hands.

Open sourcing also taps into the live brain trust of millions of people worldwide who can work with that technology, improve upon it and launch new complementary technologies or their derivatives. The cascade of innovations and new applications could result in a new economic boom. As a precedent, in the high-tech revolution which concentrated in California's Silicon Valley countless new jobs were created. That economic expansion could be dwarfed by the explosion of new business activity and uplifted living standards catalyzed by the quantum-leap energy revolution.

And instead of polluting the environment, the new technologies can be used to *remediate* existing environmental damage and geopolitical stresses, resource depletion and other major problems. So not only would an open sourced new energy era result in globally-shared material benefits, but it could relieve much of the international tensions that exist between wealthy nations and those that have long been exploited and economically impoverished.

Cost of patenting

For inventors, the patent route is a costly and difficult hurdle. For obstructionists it's an easy way to slam down the gate to the marketplace.

The patent process usually is very slow, often taking several years. The more exotic the technology, the longer it takes, especially if it involves new physics. On the other hand, open-sourcing compresses the time frame. Given the acute challenges on Earth—environmental, geopolitical, socio-economic and health concerns—does humankind have that much extra time to wait for solutions to be unleashed? Every week or month that can be saved in moving clean energy systems out into the public domain is an advantage for our civilization.

Patent protection or people protection?

An inventor from North America may obtain a global patent, but will the patent be enforceable in countries where people are dying for basic necessities?

Imagine being in a poverty-stricken country, having a family of ten and living in a village that has no electricity and no clean water and your children have no opportunity to have an education. Imagine if someone comes to your village with a new energy invention made in North America. You call together the council of elders in your village and everyone knows they can't pay the thousand dollars for that device. The whole village can't raise that amount of money. But you have a couple of very clever young people in your village who know how to tear something apart, put it back together and make it work. The invention may be a matter of life or death for those villagers. Neither inventors, their companies or governments can enforce that intellectual property protection.

Could you get a patent anyway?

Many business people and inventors assume that all valuable intellectual property must be protected by patents. Holding firmly to that mindset, they reject the open-source plan for breakthrough energy inventions as nonsense. In these cases, open sourcing advocates reply, "There's no guarantee you're going to be granted a patent. In the United States, the better your energy invention works, the better are the odds that you're not going to be issued a patent anyway. That has been the case with cold fusion discoveries, and in other arenas as well. Thousands of disruptive technologies have been denied patents."

Control the speed

Those decision makers who place their highest priority on protecting the oil industry would at best have this objection: "This open-sourcing concept wouldn't allow us to have a controlled rollout of revolutionary energy technologies. We have to have that or all hell might break loose. Events will happen that we're not going to be able to control."

The proponent of open sourcing responds, "You're not going to be able to control it anyway. None of the patents on these technologies that are launched will be enforceable; the inventions will be copied."

Will every major country promise to make certain all its citizens pay royalties for every use of a new energy technology? That's not likely. On Day Two the thing will be copied, and pirated and sent out. This already happens to products as trivial as Hollywood B-movies. It's unrealistic to think it wouldn't happen with something critical to humanity's destiny. And from the larger perspective that's beneficial, because it helps disseminate the breakthrough more quickly for the purpose of replacing polluting systems and enhancing human lives.

Even if a controlled rollout were possible, who would be the arbiters? Who are the wise ones who really know how to 'roll out' advanced energy technology and who have no vested interest? The authors would rather see the power go to the people and let it rise up from the grassroots. New wine should not be put in old bottles for very good reason; those bottles will rupture; they can't contain what's new and expansive.

Inventors' benefits

What about inventors who expect to become the Bill Gates of free energy? That too may be an impossible dream, under whatever business model is used to commercialize breakthrough energy systems. Regardless, open-sourcing does not mean the inventor will lose the opportunity for financial reward. Here's why:

If someone invents a breakthrough energy technology, develops an indisputably working model and then disseminates the knowledge around the planet, a world of possibilities opens up. The original inventor will be respected all over the world and sought, as *the* consultant. Unless the inventor wants to retire completely, he or she will be in extremely high demand to provide consulting services. A consultant can make a lot of money and in this special case the inventor would be in prime position to negotiate equity stakes in any number of start-up companies that arise from the invention and whose owners seek the inventor's technical assistance. Various companies may send invitations to be on their boards of directors, since the inventor's name will lend prestige to that company. He or she could write books, become Oprah Winfrey's friend—there are many ways the inventor will be able to profit as a result of giving away the knowledge.

There are hybrids of the open source model, all including the goal of providing material support for the inventor; inventors should be compensated for their contributions to society. Open sourcing is not a guarantee of millions of dollars for inventors, but their needs should be comfortably met at least to the financial level of an industrial scientist or mainstream academic scientist.

Open sourcing hero

Michael Couch wrote the story of an open-sourcing hydroxy hero (see chapter 14 regarding hydroxy). With Couch's permission here is a less technical adaptation from his article:[274]

> Bob Boyce was winning races in mini power boats down in Florida, using hydrogen for fuel. The problem was that hydrogen wasn't easy to locate and acquire everywhere he raced. So, he began to research making his own. Eventually he ended up making his own hydroxy on demand in what might be called the "brute force method" that drained his batteries while producing the hydroxy that fed his engines.
>
> One day he started noticing that at a certain RPM (motor speed) his power kicked up suddenly. After a few recurrences of this, he took his system apart to find out what was causing his boost at that rpm. It turned out that one of the diodes in his alternator had failed and was pumping AC (alternating current) into his electrolyzer. When that certain rpm was reached, the output from the electrolyzer suddenly increased.

> Researching further by having the gas analyzed at a local college, he found that when the frequencies kicked in he was producing much more monatomic hydrogen. He went on to perfect his electrolysis to the point that it generates a large volume of gas and mostly monatomic hydrogen.

It was a lucky accident. From the malfunctioning of a part of his system, Bob Boyce learned how to generate specific frequencies that resonate with the water molecules and cause a dramatic increase in output. Monatomic (separated atoms instead of combined atoms) hydrogen is a signature of the same type of gas that Dr. Andrija Puharich (chapter 14) was creating in his own resonant electrolyzer. So it was a breakthrough. Instead of applying for a patent and keeping his secrets to himself, Boyce has posted all the details of these experiments on the Internet for experimenters around the world. And although he is modest about what he has done, the new energy community recognizes the global ripple effect and avalanche of enthusiasm that his action is creating.

Interestingly several people whom Boyce tutored in how to produce hydroxyl later tried to patent systems that were largely based on information Boyce placed in the public domain. The authors expect that eventually they too will understand the benefits—for themselves as well as others—of opening doors by open sourcing.

20

What You Can Do

We are now able to enter into an era of abundant clean energy . . . Mankind sets the stage for a leap-forward toward a new age of civilization and culture.

—*Shin Deok-Ho, Korea,*
CEO Shinyeon Energy Research Center[275]

We're going to be able to tap into the wheel work of nature. That's the new paradigm. We will no longer have to convert precious raw materials into the energy we use for our civilization.

—*Andrew Mount, solar technologist*[276]

We must become the change we want to see in the world.

—*Mahatma Gandhi*

As climate changes and economic upheavals create a rough road ahead, humankind will need extraordinary help. Emerging quantum leap breakthroughs that tap into the free energy around us in the natural world can be powerful instruments for smoothing the road—or at least for improving humankind's prospects and helping us advance toward a higher civilization.

What does energy have to do with civilization? Everything, according to the many books predicting societal chaos alongside the end of cheap oil. However, ideas and developments you've read about in this book show that a saner future is possible as soon as the human family learns to take more responsibility for its actions.

Response-ability

Physicist and mathematician Elizabeth Rauscher earlier in her career as a science consultant was in a meeting with a group of high-ranking generals,[277] none of whom seemed worried about the ramifications of their military decisions on ordinary people or the environment. "Gentlemen," she finally asked, "don't you have *grandchildren*?"

We have similar questions for policymakers. Why is protecting the oil industry considered a matter of national security and of higher priority than the health of the upcoming generation? And who can put a dollar value on clean air?

Fortunately, at the same time that irresponsible and reckless actions are at an extreme, people all over the globe are waking up and learning. Their increasing awareness could encourage honest conversations about the biggest changeover humankind has ever faced. The change from today's world of power apartheid—a world divided into energy "haves and have-nots"—into a universal abundance of energy resources will be an unprecedented shift.

Don't underestimate people power

We, the people, must make the clean energy revolution happen. If you think that's impossible, Peter Lindemann[278] suggests looking back on how India won its independence from the British Empire. That revolution started with one man, Mohandas Gandhi, wandering around with a seemingly crazy idea that he was going to be non-cooperative with the British occupation of India. Gandhi wasn't going to hurt anybody in the meantime; he was going to bypass the situation instead of fighting the

British. Over the years he led an entire population to peacefully resist the occupation.

If someone had looked at the beginning of that movement in India and asked what the odds were of it succeeding, all the experts would have said "Zero percent probability of success." That is the prognosis given to us here, Lindemann told an audience at Reed College in Portland at the first New Energy Movement (NEM) public conference. "And they're wrong too!"

Physicist Brian O'Leary affirmed that NEM should be a *social* movement as well as a technological movement, because the problem it addresses is not lack of new energy technology but a combination of widespread ignorance, indifference, closed-mindedness, short-sighted politics, and entrenched vested interests. Dr. Steven Greer told the same conference audience that energy breakthroughs are happening in a hundred places and there are thousands of inventors coming up with more energy-converting technologies.

Researchers such as Lindemann and Thomas Valone, PhD[279] who have been examining new energy inventions for almost three decades are no longer concerned about whether "scientists who have a political agenda" agree that the new energy technology exists. Lindemann says "at first thought we tried to prove this to the scientists and show them how the 'laws' of thermodynamics don't really apply . . . . But no matter what you show them, they won't take it as proof . . . . There's no level playing field at the science level. So what we need to do is just bypass it!"

"The technology has been here, it has been suppressed, it's been reinvented and it keeps coming up because the divine forces are constantly infusing the human mind with these solutions. These solutions have to do with cooperating with nature. And living on the planet in a cooperative manner." Lindemann concludes that aggression and competition must give way to different attitudes, starting with us as individuals.

Evolution of society must indeed come before—or at least hand-in-hand with—revolutionary energy devices and systems. A weapon-making mentality won't see the gift of advanced energy as sacred. Instead of the weapons race, we look forward to increased peace-making efforts as humankind's dominant choice of reaction.

The people's world

The choice between staying wedded to polluting energy and expensive megaprojects or moving toward local energy independence profoundly affects everyone. And breakthroughs make that move possible.

Former university professor Ted C. Loder III, PhD understands the transformative potential of breakthroughs. He retired from teaching science so that he could do something even more meaningful; much of his time now is spent investigating quantum-leap energy technologies that can change the reigning paradigm and transitional technologies that can be put into use almost immediately. Most people have no idea of what is out there, he says, because all they see in the major media are people talking about solar panels, wind and biodiesel. "I think everybody in their gut realizes we aren't going to be driving our cars on the road on solar panels or windmills and growing corn to make fuel."[280]

Participants in the fledgling but fast-growing New Energy Movement[281] *are* talking about breakthroughs. They see that the brilliance of human inventiveness is sparking an explosion of solutions to global energy problems—solutions that when fully developed can empower local communities and individuals economically and also transform the environment.

The NEM organizers learned that an explosion of grassroots activism is needed in welcoming the revolutionary clean energy systems. However at this time they face the same dilemma as the quantum-leap energy technology inventors they support—lack of funding to do the job. If even a few of the movement's organizers could afford to quit their professions to be able to devote their full time to promoting the NEM message, it could quickly reach a "critical mass" of people. Writing through many nights while also attending to his day job, Joel Garbon as president of the New Energy Movement wrote draft legislation for an Energy Innovation Act, summarized in Appendix 1 of this book, but he and NEM colleague Steve Kaplan learned in the corridors of power in Washington DC that the NEM message must first reach millions of people and create mass political pressure before legislators will listen.

Breakthrough power

In addition to the sampling included in *Breakthrough Power*, numerous equally promising inventions and innovatively combined systems come to the attention of the volunteer technology assessment group, the New Energy Congress,[282] led by directory Sterling Allen. On the New Energy Congress you can view a regularly updated database of promising energy technologies as well as a daily web newsletter. As this is written, today's

Free Energy Updates newsletter carries news from Korea.[283] A research center in Korea [284] says it has developed two highly exotic prototypes. One is a piston-based permanent magnet motor that the inventor claims doesn't require fuel or input of electricity; it's powered by magnetic attraction and repulsion instead. The other breakthrough is a novel wind generator that allegedly puts out megawatt power by harnessing engineered differences in air pressure. Its top part contains a 30-ton propeller, levitated on magnetic bearings and looks like a flying saucer spinning on top of a solidly built tower.

In China, another inventor claims to have energy inventions powered by cosmic energy. A Japanese company announced a car that runs on water. Time will reveal how many of these paradigm-breakers will succeed on their journey to the marketplace. Meanwhile an array of breakthroughs in less exotic clean energy systems are ready for use by the human family.

Like a flyweight facing a heavyweight wrestler, none of these new energy companies can confront the combined oil, gas, coal and nuclear fission related industries and say "Step aside, buddy; you've messed up my planet." Rather than confrontation, some of the new energy entrepreneurs are finding other paths past the vested interests such as looking for niche markets that the energy establishment is not serving. Why fight the giant, especially in its territory? These new energy entrepreneurs are wisely focusing on creating new structures and networks rather than antagonism.

New energy researchers and inventors are creating their own opportunities, sometimes in other countries where the vested interests have less of a presence. It is more important to move on to positive actions without stopping to wallow in anger about injustices—including the unjust suppression of inventors. Exposing the skeletons in the closet serves to enlighten, but getting off-message with retribution will be counter-productive.

Jean Houston is an author who has taught people about how to renew self and society. She points to history to show that this is possible under any circumstances. During the years 1400 to 1500 the men and women of the warring and disaster-plagued Italian Renaissance did not give up. Despite fire, flood, plagues and personal loss of every kind, they propelled humankind into the modern era, meanwhile celebrating life through art and other expressions of beauty.

"We live in strikingly similar times, and we need much the same quickening charge to help us become stewards of a time in history in which we will either devastate the planet and die as a species, or grow mentally and spiritually into our Renaissance selves, able to take on the challenge and tasks of this most potent and dangerous era."[285]

The human family does face danger. The scientist from NASA whom we met in chapter two, Dennis M. Bushnell PhD, stresses the urgency of the need to stop the use of oil and coal. When global warming and melting ice releases methane from oceans and the Arctic, he says, a major tipping point is possible and probable. Massive methane releases in the Earth's past history resulted in extinctions of up to eighty per cent of species at those times.

Yet the polluting energy industries today hold geopolitical power and don't intend to hand their government subsidies over to the clean energy sector. And government officials lack the imagination so far to figure out another source of income instead of fuel taxes.

This multi-faceted dilemma could galvanize the human family instead of causing it to give up and resign itself to becoming extinct. The authors believe the human spirit is capable of more courage, creativity and cooperation than we realize. That was demonstrated to Manning during crises such as a forest firestorm which swept the outskirts of a city in British Columbia. She and ten thousand other residents were suddenly evacuated from their homes, and as a result many ordinary people became heroes and heroines rising to the challenges of that fiery time.

On a larger scale the human family could work together heroically when enough people realize that we are heading into an atmospheric-pollution crisis. The brightest minds and hearts on Earth could meet together in emergency sessions outside the oil-soaked old political structures, agree on an action plan and mobilize it.

Start locally

Through whatever skills you have, you can help change the dominant paradigm of what and how people see. For instance, Bruce Sterling is an artist whose theme has been the impact of technology on society.[286] He noticed that atmospheric pollution is only partly a political problem; at its core it's a consumer problem and a design problem. As an artist he saw an aesthetic issue; people don't even notice the toxic exhaust emissions in the air they breathe. "We're in trouble because we live in filth and we can't see it. We're like eighteenth century people who lived before germ theory. We're ignorant of the squalor that surrounds us, and we have bad taste."[287] Designers can help change that. As an example he advised anyone who is trying to sell electric cars to give away smog detectors. They could install smog sensors and the software to display the information in every dashboard. "Make the invisible visible. Let people see. The rest will follow."

New energy advocates with important ideas to convey can learn from communicators such as Sterling who present their messages with wit and style and appeal to the human senses. A helpful primer for improving the chances of worthy ideas being heard is the book *Made to Stick: Why Some Ideas Survive and Others Die*.[288] Its authors wryly quote Tesla's friend Mark Twain who once observed, "A lie can get halfway around the world before the truth can even get its boots on."

Grassroots groups such as NEM can help share the truth about the abundance of clean energy choices and make the invisible visible by supplying new energy information to the media, schools and colleges, governmental representatives and agencies and others. Such efforts build the critical mass needed to bring quantum-leap solutions to the energy crises to public attention. Consider setting up a discussion group in your area to educate your friends and neighbors on new energy issues and to help them see this as an opportunity to work together for a brighter future. They may be thinking that everything is taken care of because green energy is so often in the news. Your associates may not realize that it takes breakthroughs and a grassroots push to support their development and widespread use. Otherwise, clean energy technologies will continue to be incrementally improved and will merely supplement polluting energy technologies, not replace them.

Following are more suggestions for how you can help create a welcoming culture for an energy shift. You can:

Learn and help others learn

- Subscribe to a new energy publication or inform yourself online (see Resources section of *Breakthrough Power*).
- Start a discussion group with positive goals.
- Help organize a New Energy Movement series of informative meetings.
- Learn more about creative processes that motivate groups to *work together* for sustainable solutions.[289]

- Invite your neighbors/ associates to look at breakthrough energy solutions.
- Fire up the enthusiasm of an inventive teenager.
- Arrange for you, or a new energy educated speaker whom you know, to be invited to visit a school classroom.
- Give copies of *Breakthrough Power* as gifts to the women, men and youths in your life.
- Gift your local libraries with copies of *Breakthrough Power*.

Alert decision-makers

- Speak (politely and factually) to your elected officials; tell them they have options other than carbon fuels and nuclear power plants.
- Be blunt with evasive politicians and tell them "Get on board by acknowledging the need for breakthroughs, not just incremental improvements to clean energy alternatives. Support and clear a path for them or we're not voting for you!"
- Vote in elections for progressive supporters of powerful clean energy alternatives.
- Form neighborhood and regional networks.

Financially help support the efforts of new energy inventors and *educators*

- Encourage financial help for harmonious energy research.
- Seek a philanthropist's support for it.
- Support the New Energy Movement in whatever way you can.

Begin with self, to promote a world that's safer for breakthroughs

- Focus on what people all have in common.
- Promote unity among people with different views.
- Notice your own fear of change.
- Promote healing laughter.
- Promote cooperation and compassion in everyday life.
- Forgive the stonewalling scientists, bureaucrats and other human obstacles.
- Move on to positive plans and actions.

Imagine a better world

- Remind yourself that a vision gives more light than a frustrated rant.

- Find your own best path toward the personal and societal renaissance.
- Find friends and allies who support your positive direction.

Wade Frazier[290] was deeply involved in an effort to bring alternative energy to the American marketplace in the 1980s. Through the years he has heard two prominent motivations for making free, forever renewable, non-polluting energy a daily reality for humanity: On the upside, the positive motivation is that an unprecedented human journey based on abundance instead of scarcity can be catalyzed by a transition to free energy, while healing the environment at the same time. The downside motivation is to avoid catastrophe; today's energy production methods are destroying the biosphere, energy scarcity results in poverty, and the Middle East is increasingly the site of energy wars which may ignite World War III. Frazier advocates envisioning the opposite.

> Abundance can become humanity's reality, but we first have to imagine what it would look like. Do we want to pursue abundance? The choice is ours.

Setting the tone for a new era

Choosing the power of love over love of power could be the first step in the grassroots push toward wise choices and an eventual civilization of peace and prosperity based on energy abundance. While organizing the first New Energy Movement (NEM) public conference,[291] co-author Joel Garbon asked singer Shawn Gallaway to open the event with a particularly inspiring song from Gallaway's repertoire. When they heard it, the New Energy Movement board of directors decided unanimously that the song "embodies the energy from which this movement springs." Following is an excerpt from that song, "I Choose Love."[292]

> *I see us dawning as one world united—so what do you choose?*
> *Love or fear? We choose.*
> *And I choose to feel the whole world crying*
> *I choose to hear the one voice rising*
> *I choose to feel a song united in the strength that we can rise above*
> *I choose Love!*

The human family has a rainbow of choices and the opportunity to progress to a peaceful higher civilization. Your awareness of the issues and attitudes surrounding new energy systems can help fellow citizens on that path toward that better world. Breakthrough inventions are a powerful tool for elevating economies and prying society away from the technologies that produce oil slicks and radioactive wastes, but they are merely a technological advance unless attitudes of justice and sharing prevail on Earth.

You have the opportunity to add your caring, your voice, your actions and power of visioning to the emerging movement of planetary citizens toward wise choices. Please join in creating that tidal wave of wisdom.

APPENDIX 1

Energy Innovation Act draft legislation:

written for USA but adaptable elsewhere

The following is the first part of a draft presented to various members of the United States Congress and their legislative assistants by the executive staff of The New Energy Movement (www.NewEnergyMovement.org see link on the home page) in 2007. It has not yet found a sponsoring legislator.

While this draft legislation concerns support for accelerated research and development of "new and unconventional" clean energy technologies, the New Energy Movement at the same time supports the role of conventional renewable energy technologies (wind, solar, biofuels, hydroelectric) and welcome advances in conventional energy technologies such as in thin-film solar photovoltaics and new wind turbine designs. However, the well-known conventional renewables have the advantage of ready access to a growing pool of market capital and government resources to finance R&D. In contrast, unconventional energy technologies are generally little-known and poorly funded, yet have the potential to produce the quantum leap breakthroughs in clean energy generation so desperately needed in this age.

A publicly viewable database of new energy technologies, both conventional and unconventional, is regularly updated by the organization New Energy Congress and can be viewed at www.NewEnergyCongress.org, "Top 100".

The draft legislation has a provision for establishing a new Office of Energy

Innovation as a Joint Congressional Office performing the type of activities normally reserved for executive branch agencies. This unusual provision focuses the debate on the need for critical new approaches to serious and growing energy-related problems. Specifically, the new Office should not be established as an agency of the Department of Energy, which has a well-documented history of vested-interest obstructing breakthrough energy technologies. To paraphrase a well-known saying, "New wine should not be put in old bottles.")

ENERGY INNOVATION ACT OF 2007 (DRAFT)

Preamble

The United States is faced with unprecedented challenges to its national security, environment, and economic growth and stability due to a combination of factors that relate directly and indirectly to energy policy. (See the full draft legislation at www.newenergymovement.org for a full listing of these factors and their connection to energy policy. The list was omitted here.)

Despite some modest progress in the adoption of conventional clean and renewable energy sources such as wind power, solar power, and biofuels, these technologies have significant limitations and can satisfy only a small fraction of the growing U.S. demand for energy. Based on current U.S. energy policy, the Energy Information Administration (the official statistics and projections agency of the Department of Energy) projects that by the year 2030 domestic energy consumption will be 30% higher than 2005 levels, with fossil fuels supplying 86% of the total energy demand, leading to a 35% increase in carbon dioxide emissions. Petroleum consumption is projected to be 20% higher than 2005 levels, with imports providing 60% of the supply. Despite the array of energy-related challenges facing our nation and planet, these projections suggest that current policies will leave the country in a precarious and highly-vulnerable state even 25 years from now.

America must embark on a bold new path with serious commitment and urgency.

The single most highly-leveraged opportunity for advancement toward solving complex global problems lies in a transformation in the way humanity generates and uses energy. The discovery and widespread deployment of advanced clean energy generation systems can lead to a future of breathtaking promise and near-limitless possibilities for the United States and the greater global community. A partial vision of the possibilities includes:

- Mitigation of global warming and stabilization of climate patterns.
- Clear, clean, healthy air through elimination of air pollution from industrial sites and vehicle fleets. Mercury contamination and other forms of toxic pollution from combustion of fossil fuels become a historical footnote.
- Cessation of military conflicts and geopolitical tensions related to dwindling fossil fuel supplies and other natural resources, resulting in greatly enhanced national security, reduced U.S. military expenditures, and reduced risk to military personnel and civilians.
- Reduction of the proliferation of potentially destructive nuclear technologies.
- Reduction or elimination of the generation of long-lived radioactive wastes.
- Mitigation and remediation of water and soil pollution made possible by very low cost energy.
- Great reduction in environmentally-damaging resource extraction and transportation of fossil and nuclear fuels.
- Restoration and preservation of forests formerly depleted for wood fuel.
- Sustainable low-impact forest management made possible by selective logging with aircraft powered by inexpensive new energy technology, eliminating the need for logging roads and preserving stream quality.
- Cessation of environmentally-damaging hydroelectric dam building and removal of dams that are devastating to fish runs and which inhibit the natural cyclical replenishment of flood plain soils.
- Greatly increased recycling of wastes made possible by very low cost energy, further reducing the need for environmentally-damaging resource extraction.
- Decentralization of energy generation eliminates the vulnerability of a centralized electrical grid system. Expensive, dangerous, landscape-altering power transmission lines become obsolete.
- Hunger and thirst disappear as food can be grown in compact automated systems anywhere on the planet, and water can be extracted from the air and sea and repeatedly recycled using low cost energy.

- Global standard of living greatly improves using clean inexpensive energy to foster sustainable local economies.
- Education rates improve throughout the developing world as subsistence living conditions are eliminated.
- Human population stabilizes as third world birth rates decline, the result of an educated global populace, thriving local economies, and higher living standards.
- A great expansion of the U.S. and global economy occurs as new industries and businesses are spawned by advances in energy generation.
- A new era of space travel is ushered in with the development of advanced energy and propulsion systems.
- A tremendous wave of human creativity is unleashed as people are freed from toil for basic sustenance, producing unimaginable progress in social and material conditions. A global culture of sharing and cooperation is embraced.
- The possibility for true and lasting world peace is within humanity's grasp.

Summary of the Energy Innovation Act of 2007

The Energy Innovation Act of 2007 contains the following provisions:

- Establishes and funds a new Joint Congressional Office of Energy Innovation

Assigns to the Office of Energy Innovation the following mission:

- Identify and rapidly advance new and unconventional approaches to energy generation in recognition of their critical importance to the United States' national and energy security and their potential as valuable solutions to urgent global environmental and resource depletion issues.
- Accelerate small business early-stage research and development of new and unconventional approaches to energy generation through awards of modified Small Business Innovative Research (SBIR) grants and loan guarantees designed to fund up to five years of R&D.
- Award large cash prizes for outstanding achievements in new and unconventional approaches to energy generation.
- Perform initial and biannual technical assessments of candidate

technologies through collaboration with contract consultants and various qualified university, private, and national laboratories.
- Track progress of candidate technologies through annual progress reports.
- Engage the American public's enthusiasm and support for new energy options through high-profile publication of award recipients, general descriptions of the technologies, and nonproprietary summaries of technical assessments and annual progress reports.
- Establish regional research centers and incentives for collaborative work among researchers expert in a particular energy technology category.
- Administer an Energy Innovation Fund to receive royalties from the profits of commercially successful federally-supported technologies, and use these royalties to fund ongoing future grants.
- Collaborate with the Department of Energy on strategies for widespread deployment of viable commercial technologies.
- Establishes an independent and publicly accountable Citizen Oversight Council to monitor the Office of Energy Innovation and ensure compliance with its stated mission.

Authorizes appropriations of $2.25 billion dollars per year for ten years, beginning in 2007, as follows:

- $2 billion/year for awards of modified SBIR grants, loan guarantees, and cash prizes
- $200 million/year for initial and biannual technical assessments of candidate technologies
- $50 million/year for facilities, administrative staff and overhead of the Office of Energy Innovation

It is expected that the federal funding awarded for two to five years of research and development of promising new energy generation technologies will result in many such technologies advancing to the stage of demonstrating substantial commercial potential, at which point private investment capital will provide further funding to the technologies of merit.

The economic stimulus and resultant federal, state, and local tax revenues generated by the provisions called for in the Act are substantial. Successful commercial ventures arising from the federal funding and "incubation" of promising concepts/technologies will multiply the economic stimulus and tax revenues many times over through new enterprise, creation of new jobs, and a general expansion of the U.S. economy that will rival or surpass the impact of the computer revolution.

The per capita investment by American citizens in the provisions called for by the Act amounts to less than $8 per year. The potential returns as measured by positive impacts to national security, energy security, human health, the environment, and economics are enormous and incalculable.

APPENDIX 2

Mohorn's Self-running Wall Dryer

This invention is neither a generator of electricity nor a fuel. However it is a breakthrough invention that taps into a previously-unrecognized form of energy to do useful work with no batteries, no chemicals and no maintenance needed.[293]

Austrian engineer Wilhelm Mohorn[294] invented an unusual energy-converting device that demoisturizes damp walls or buildings without the use of standard electrical current—or any conventionally recognized energy input. His invention has already saved old cultural monuments from decay.

European buildings are often made of brick or sandstone. In such buildings particularly, the natural capillary action of water lifts underground dampness up onto basement walls. Drying out those wet basements has traditionally been a huge costly energy-consuming problem. Mohorn's invention doesn't work where water in a basement comes from flooding, such as a broken water main. It does however work demonstrably in the countless wet basements where the problem is caused by capillary action of water.

Mohorn has received honors and awards for his invention. In 1995 he was given Austria's state prize, the Kaplan Medal. More than 38,000 of his "Aquapol" units have been installed successfully in Europe, including in the Hungarian Parliament in Budapest and in the Haydn Museum and the Castle Reitenau in Austria.

As so often happens with the new paradigm, today's science can't explain exactly why and how his device works. The inventor believes that energy from the all-permeating background (zero point energy) seems to play an important role in it.

The story began when Wilhelm Mohorn was a student, playing percussion in a band. His instruments were kept in a damp cellar and got increasingly rusty. This not only annoyed the hobby percussionist, but also spurred him to think about the problem of damp walls—and look for a solution. The problem is that conventional devices that dry out damp walls consume large amounts of electrical current when in continuous operation, and also tend to break down. Inventive, Mohorn began to experiment.

From his study of physics Mohorn knew that water molecules can be directed with certain energy or force–fields. After doing experiments, he further assumed there might be a dynamic earth-force-field. If it existed, would it be possible to use this physical energy-field? To test his hypothesis, he began to develop cylindrical air-core chokes—coils without iron cores—which he thought should receive such a field.

Coils alone didn't show any effect, so he got the idea of turning their ends into antennae. After arranging the coils in a certain order, he put them in the cellar. Soon the musty smell disappeared. With the help of measuring instruments for monitoring air humidity and the dampness of walls, he observed a dehydrating effect. He didn't know exactly why, but his device was working.

Soon the young inventor expanded his experiments into the homes of any friends who complained about damp walls. He would install his device, placed within a woven box that looked somewhat like a beehive, and to everyone's surprise damp cellars drained noticeably after months. In one case an extremely damp floor dried out—again without using electrical current.

After two years of testing, Mohorn applied for the first patent on his device which he called Aquapol. The name is from the Latin *aqua*, water, and *pol*, short for polarization. He called the physical principle of his discovery magnetokinesis, because with the Aquapol devices a magnetic-like field causes humidity to move.

After the patent was granted, Mohorn founded the company Aquapol in 1985. He continued to improve his invention and apply for more patents. It was getting easier to convince people about the efficiency of his technology. The German people whom Manning spoke with said Mohorn's Aquapol system eventually became a thorn in the side of the construction industry in his region, because the industry lost income.

Renovation specialists had had an ongoing source of income from repairing the repeated encroachment of plaster which is otherwise common with damp old buildings.

There were setbacks as well as successes; some customers complained that the installed device did not work with them. Mohorn agreed that his technique had in those cases failed, and he refunded their money. He says he despaired until he recognized the causes of those cases. Near radio and TV broadcasting and other broadcasting stations, his antenna-like units were disturbed and could not work as intended. He solved the problem by building electromagnetic interference suppressors for his devices.

Sometimes there were other unclear cases in which the device didn't seem to work. Mohorn found no explanation until he heard the late Shiuji Inomato, PhD physicist from Japan, lecture in Hungary. After being inspired by Inomato, Mohorn felt that he would find the solution for his technical problems in the area of space energy—zero point energy.

For two years the inventor backed away from his company to investigate the possible connection of such energy with his Aquapol device. Gradually he developed a theory of *why* his units worked only half the time in certain places. In his view it was because the dynamic earth-force-field constantly changes its polarization at some places. He rebuilt his device so it adapts to the changing polarization, and he developed a new reception antenna which worked nearly everywhere.

Another of his problems was solved during this time. Previously, after the installed devices were cleaned and reinstalled they often had ceased to work. After much testing he discovered how to make the device work independently of its direction.

How exactly are Aquapol devices made? That information is kept proprietory by Mohorn. We do know that the devices consist of three main parts: the receiver unit, the polarization unit, and the broadcasting unit. The receiver unit is made of two different flat spiral antennas. Mohorn believes the device sucks in the "dynamic earth-force-field" which becomes funnel-shaped at that location. The polarization unit consists of a cylindrical air-core choke that continuously polarizes the received energy, which is right-circling. This polarised earth-force-field is passed on to the broadcasting unit. The broadcasting unit consists of special antennas and three tetrahedron-shaped redirection-coils which are also cylindrical air-core chokes. This broadcasting unit transmits the polarized earth-force-field into the room and causes water molecules in damp walls to "walk" downwards.

Until 1991 Mohorn thought that the supposed earth-force-field alone made his units work. But he found that the radius of activity built up by the Aquapol-device is in practice much larger than the conical energy-suction-space. After hearing Dr. Inomata's lecture, Mohorn speculates that space energy flows from above into the unit to reinforce it. He says that in 1995 this was proven with tests in Graz, Austria. We do not have details on those tests, however.

Converting background free energy

Mohorn seems to have discovered a totally new type of generator which changes space (zero point) energy directly into another form of energy. According to the inventor, the energy-form used by his device has a wave character similar to known electromagnetic waves. But he believes there's a difference in how the wave structure in his devices is composed and built. In his view, the new-found wave type consists of a magnetic wave and a gravitory wave which circles around the magnetic wave. Both wave components circle around the spreading axis. The new-found wave operates differently than electromagnetic waves as emitted from satellites for example. Mohorn calls this newly discovered energy-form 'gravomagnetism' and says it is similar to electromagnetism. As the term indicates, he is talking about a combination of gravity and magnetism.

"In my opinion," Mohorn says, "gravity actually is a product of space energy which penetrates earth. Out of the difference of the space energy which enters our planet and that which emerges from it again, a force arises which is directed towards the earth . . . . According to my findings the magnetism of earth results from the rotation movement of our planet rather than from ferromagnetic matter within the earth. The Swiss physicist Christian Monstein was able to prove this experimentally and mathematically. I even go one step further and claim that the earth is rotating because space energy enters increasingly and is funnel-shaped at its poles. If entering space energy causes the earth to rotate, and this rotation causes the magnetism of earth, the conclusion is that also the magnetism of earth is a product of space energy."

Mohorn is convinced that all energy forms known to us are a manifestation of space energy. "When it (cosmic energy) interacts with matter or penetrates it, it partly changes into different forms of energy—gravity, magnetism, gravomagnetism etc. When it penetrates the earth layers it is marked by every material with its matter-specific frequency. Thereby the gravomagnetic earth-field with its different frequencies is built. At the earth's surface it emerges in reduced form. And it is exactly this low-powered earth-field which enters my Aquapol-device from below, and it is reinforced by entering space energy from above."

Mohorn's dehydration units only emit or broadcast right-spinning force-fields. The inventor is convinced that in the future these biologically-active fields will also be used for healing purposes. In experiments, however, the Austrian engineer also managed to broadcast left-spinning force-fields. He noticed those fields had the opposite effect, causing humidity to be moved upwards. This effect might perhaps be useful in agriculture to moisten dry soil during droughts. "One of my further goals is to also fertilize deserts with this technology," says Mohorn.

Mohorn's theories oppose conventional science's teachings about earth's magnetism and gravity. Regardless, the Aquapol-device works. It saves historic buildings from the destructive effect of wall dampness, and users only pay a fraction of the costs that they would spend for conventional drainage methods. Mohorn's units drain excess moisture out of buildings in an environment-friendly manner and without electrical current. In the late 1990s he calculated that his 20,000 installed Aquapol units in Europe help save enough megawatts of electrical current to heat about 2,000 average houses or medium-sized apartments. Recently the German physicist Thorsten Ludwig, PhD, told us that 38,000 Aquapol wall dryers have sold, each with a guarantee.

The unusual wall-dryer survived the anger of the construction industry and challenges to its authenticity—challenges raised by skeptics and his competitors. "The results have even been investigated by court cases," Dr. Ludwig reports. A public examiner did the testing and found the Aquapol invention does work.

German engineer and publisher Martin Meier told Manning that the Austrian inventor has set himself an ambitious long-term goal: not only saving electrical current and heating energy but also producing energy. But for now Mohorn is happy that, thanks to his invention, his little castle in Reichenau on the Rax river that was once very damp now has dry feet again. From what Manning heard in Europe, Wilhelm Mohorn is a gifted intuitive whose non-conventional but effective inventions raise further questions about the mysteries of water and of zero point energy.

NOTES

1. William Rees of the University of British Columbia coined the phrase "ecological footprint" to show how much productive land and sea area is needed to regenerate the resources a person or human population consumes.
2. Hilscher, Gottfried. October 5, 1995 letter to author.
3. April 27, 1984, letter to Bill Muller from Zinsser, West Germany.
4. Centimetre is the European and British spelling of centimeter.
5. Www.mullerpower.com.
6. Early in their lives, Carmen and her brother Billie's surnames were changed from Muller to Miller.
7. Brian O'Leary, PhD, in Foreword to *The Coming Energy Revolution.*
8. Hawken, Paul, *Blessed Unrest: How the Largest Movement in the World Came Into Being and Why No One Saw It Coming*, Viking Press NY 2007.
9. Newenergymovement.org and newenergymovement.ca.
10. Professor emeritus John O' Malley Bockris to co-author Manning.
11. Preschool children's rate of asthma rose by 160 per cent between 1980 and 1994—more than twice the national average according to a report by Harvard University researchers and the American Health Association April 29, 2004 report *Inside the Greenhouse: The Impacts of CO_2 (carbon dioxide) and Climate Change on Public Health in the Inner City.*
12. Lenz's Law about equal-and-opposite reactions.
13. Felenstein won the 2007 ACE award.
14. Brian Fuller, editor of E.E. Times.
15. Dr. Hermann Scheer has received many awards, including the World Solar Prize, the World Wind Energy Award, the World Prize for BioEnergy and the Alternative Nobel Prize.
16. *Energy Autonomy: The Economic, Social and Technological Case For Renewable Energy*, Hermann Scheer, Earthscan, Sterling VA, 2007, page 3.
17. See Dr. Helen Caldicott's book *Nuclear Power Is Not The Answer*, 2006, The New Press, London and New York.
18. Caldicott, Helen *Nuclear Power is Not the Answer*, The New Press, NY 2006, page 126.
19. Limestone is calcium carbonate.
20. Lime is calcium oxide.
21. Italicized emphasis added.
22. *Energy Challenges: The Next Thousand Years*, March 30–April 2, 2007, sponsored by Foundation for the Future, Bellevue, Washington.
23. Russ George, PhD.

24. David Houle's May 29, 2007 interview with Dennis Bushnell, www.evolutionshift.com.
25. New Energy Movement conference, Portland, Oregon, 2004.
26. *Aquarius Now: Radical Common Sense and Reclaiming Our Personal Sovereignty*, p. 40 Excerpted from Aquarius Now by Marilyn Ferguson with permission of Red Wheel Weiser. 1-800-423-7087 www.redwheelweiser.com.
27. South Pacific musical.
28. William Tiller, PhD, *Science and Human Transformation: Subtle Energies, Intentionality and Consciousness*, Paviour Publishing, Walnut Creek CA, 1997.
29. Www.theintentionexperiment.com.
30. There are abundant nonpolluting solutions to soil problems, such as remineralization with rock dust. The opposite problem of salt-encrusted lands can also be tackled.
31. Tiller explained, to an Institute for New Energy audience, that the shaman's consciousness was broader than the other islanders.
32. Davidson, Dan A. *Shape Power: A Treatise on How Form Converts Universal Aether Into Electromagnetic and Gravitic Forces*, Rivas, Arizona, 1997.
33. Davidson, Dan A. *Energy: Free Energy, Aether and Electrification*, Sierra Vista AZ: RIVAS, 1992.
34. McKie gives credit, for the ideas about the city, to the late Frank Lloyd Wright's 1932 low density "Broadacre" city concepts. Broadacre is now controversial in architects' circles, but one defender says Americans got the suburban sprawl shell of Broadacre City without the social reform substance.
35. U.S. Department of Defense Program Solicitation for Fiscal Year 1986, AF86-77 Non Conventional Propulsion Concepts.
36. *Lost Science*, Gerry Vassilatos, Adventures Unlimited Press 1999 p. 106.
37. Cheney, Margaret, *Tesla: Man Out of Time*, Laurel, NY 1981 p. 39.
38. More information on the difference between open-path systems in contrast to closed-path systems is available from www.free-energy.ws, Dr. Peter Lindemann's DVD, *Tesla's Radiant Energy*.
39. William Stanley invented the induction coil, U.S. patent #349,611, a transformer that creates alternating current electricity. In 1886 he built the first AC system, providing lighting for offices and stores in the town of Great Barrington, Massachusett. His company made transformers, auxiliary electrical equipment, and electrical appliances. In 1903 General Electric bought the Stanley Electric Manufacturing Company.
40. *The Free Energy Secrets of Cold Electricity*, Peter Lindemann, Clear Tech Inc., 2001.
41. Tesla's Big Mistake? by William J. Beaty, ExtraOrdinary Technology magazine, Jul/Aug./Sept. 2007, p 19–22.
42. Http://web.mit.edu/newsoffice/2007/wireless-0607.html.
43. "Cold electricity" motor of the late Edwin Gray, see *The Free Energy Secrets of Cold Electricity*, Peter Lindemann, Clear Tech Inc., 2001.
44. King, Moray, *The Energy Machine of T. Henry Moray: Zero-Point Energy and Pulsed Plasma Physics*, Adventures Unlimited Press, Kempton, IL, 2005, p. 21.
45. Tesla used the term Radiant Energy to describe a strange cold-current form of

electricity, and Moray adopted the term. We retain the capital letters to differentiate the concept from standard radiant energy.

46. Moray, T. Henry and John Moray, *The Sea of Energy In Which the Earth Floats*, Cosray Research Institute, Salt Lake City, 1978.
47. A footnote in *The Sea of Energy* says this is what led Henry to his research in semiconductive materials.
48. King, Moray, *The Energy Machine of T. Henry Moray: Zero-Point Energy and Pulsed Plasma Physics*, Adventures Unlimited Press, Kempton, IL, 2005, p. 13.
49. Writing in the book *The Sea of Energy.*
50. King, Moray, *The Energy Machine of T. Henry Moray: Zero-Point Energy and Pulsed Plasma Physics*, Adventures Unlimited Press, Kempton, IL, 2005, p. 17.
51. That hypothesis includes "abrupt, synchronous ion surges in plasma appear to coherently activate the zero-point energy." [King, Moray, *The Energy Machine of T. Henry Moray: Zero-Point Energy and Pulsed Plasma Physics*, Adventures Unlimited Press, Kempton, IL, 2005, p. 17.]
52. Schauberger quoted by Callum Coats, *Living Energies*. Gateway Books, Bath England 1996, p. 4.
53. Dr.Hartmut Muller, "The Universe's Energy Source", Raum & zeit Special, 158 page report, Wolfratshausen, 2002, p. 56.
54. The science of hydraulics deals with practical uses for liquids in motion.
55. Schauberger quoted by Alick Bartholomew in *Hidden Nature*, Adventures Unlimited Press, Kempton, IL, 2005 p. 15.
56. Nick Cook, *The Hunt for Zero Point*, Random House, 2001, p. 211.
57. Nick Cook *The Hunt for Zero Point.*
58. A series of books by Australian author Callum Coats covers Schauberger's knowledge extensively. *Hidden Nature* by Alick Bartholomew introduces the concepts for the general reader.
59. Austrian patent # 117,749.
60. William Baumgartner www.vortex-science.com.
61. Russell, Walter, "An Open Letter to the World of Science", from Russell's book *A New Concept of the Universe.*
62. David, *Wilhelm Reich: The Evolution of his Work*, Arkana, London, 1985, p. viii.
63. Pond's website is www.svpvril.com.
64. Glenn Clark, *The Man Who Tapped The Secrets of the Universe*, The University of Science and Philosophy, Waynesboro Virginia, 1953, p.4.
65. Walter Russell was so far ahead of his time that in 1926 he predicted the existence and characteristics of tritium, deuterium, neptunium, plutonium, and other elements that weren't discovered until the 1930s and 1940s. His Periodic Table of Elements was unusual—spiraling and related to musical octaves. Carbon is in the middle of Russell's table, and the ends extend beyond the standard table of elements to pre-hydrogen in the "generative" direction and in the opposite "degenerative" direction beyond the known radioactive elements. Today's scientists have no instruments to detect those pre-hydrogen elements, but Russell's students believe that is where discoveries in harnessing beneficial cosmic energy will be made.
66. Detective work by electrical engineer Toby Grotz made the documents' existence public.
67. Tim Binder's speech, Institute for New Energy Symposium 1994.

68. Www.aetherometry.com.
69. Http://free-energy.ws/.
70. Hyypia, Jorma "Amazing Magnet-Powered Motor" Popular Mechanics spring 1980.
71. U.S. Patent No. 4,151,431.
72. Sweet, Floyd A. "The Vacuum Triode Amplifier". In *Free Energy: Final Solutions*, ed. Don Kelly, Clearwater, FL, 1990.
73. Sweet, Floyd A., Nothing Is Something: The Theory and Operation of a Phase-Conjugated Vacuum Triode. 24 June 1998.
74. Sweet, Floyd, and Thomas Bearden, "Utilizing Scalar Electromagnetics to Tap Vacuum Energy". In *Proceedings of the 26th Intersociety Energy Conversion Conference*. In Boston, 4–9 August 1991, LaGrange, Il: American Nuclear Society, 1991, Vol. 4 No. 1:370–375.
75. The field of phase conjugate optics.
76. Sweet, Floyd, and Thomas Bearden, "Utilizing Scalar Electromagnetics to Tap Vacuum Energy". In *Proceedings of the 26th Intersociety Energy Conversion Conference*. In Boston, 4–9 August 1991, LaGrange, Il: American Nuclear Society, 1991, Vol. 4 No.1:370–375.
77. Floyd Sweet to Jeane Manning.
78. Www.cheniere.org.
79. U.S. Patent 10,239 "Improvement on Magneto Electric Machines, July 16, 1879.
80. Www.rexresearch.com/gary/gary1.htm.
81. The late Ervin Krieger followed the paper trail in Britain before files and a model disappeared. He told Manning that the Coler device had contained a high-frequency component that was beyond the measuring capacity of German instruments.
82. The late Ervin Krieger to Manning.
83. R. Hurst, Ministry of Supply; and Captain R. Sandberg, Norwegian Army.
84. As quoted in "Hans Coler's Free-Energy Generators", Robert Nelson, *Nexus* magazine October-November 1998.
85. *British Intelligence Objectives Subcommittee report: B.I.O.S. Final Report No. 1043: Item No. 31* "The Invention of Hans Coler, Relating to an Alleged New Source of Power".
86. Tom Valentine, "EMS_Electronic Power That Could Change the World's Economic Power Picture," defunct *Newsreal* Series magazine.
87. Capacitors store and release electric charge.
88. Tom Valentine, Ibid. p. 37.
89. Valentine, Ibid. p. 38.
90. Ibid.
91. Lindemann, Peter, *Free Energy Secrets of Cold Electricity*, Clear Tech Inc.; 2001, p. 13.
92. "Normally, electricity consists of positive and negative particles, but Gray's system is capable of using one or the other separately and effectively," according to Tom Valentine. See book by Peter Lindemann, *Free Energy Secrets of Cold Electricity*, Clear Tech Inc.; 2001,and Thomas Bearden's www.cherniere.org.
93. Gray's machine is analyzed in the DVD documentary series "Energy from the Vacuum". See www.cheniere.org.
94. see book by Howard Johnson, *The Secret World of Magnets*, Cheniere Press; and DVD 4 in the Energy From the Vacuum series, www.cheniere.org.

95. See DVD "Electric Motor Secrets" by Dr. Peter Lindemann, www.free-energie.ws.
96. Blue Energy Canada website.
97. Former U.S. President John F. Kennedy as read into Congressional Record, by Hon. Thomas J. Dodd; concerning the Passamaquoddy Tidal Power Project; as reported in *Congressional Record*–Appendix; October 22, 1963; pp 6580–6581. [Proceedings and Debates of the 88th Congress, First Session.]
98. Excerpts of address by Senator John F. Kennedy, Democrat, Mass., concerning the Passamaquoddy Tidal Power Project ; at Maine Democratic Party issues conference banquet, Augusta, Maine, Sunday November 15, 1959; as reported in *Congressional Record*, September 1, 1960; p. 19514.
99. Barbara A.F. DelloRusso, "Waves of the Future? Harnessing Tidal Power, *Infinite Energy*, Issue 33, 2000 p 44–46.
100. Michael Maser of Blue Energy Canada.
101. DelloRusso, ibid.
102. Larrabee, D.R. 1973, "Tidal Power? France Wouldn't Try it Again, *Maine Sunday Telegram*, June 17, 1973: sources for Barbara DelloRusso in *Infinite Energy*, Issue 33.
103. Writer for the *Maine Sunday Telegram*.
104. DelloRusso, Ibid.
105. Www.dreamofpassamaquoddy.com.
106. 2004 New Energy Movement conference, Portland, Oregon.
107. Valone, Thomas, *Zero Point Energy: The Fuel of the Future*. Integrity Research Institute, Maryland, 2007, p. 10.
108. Www.energy-science.org.uk.
109. Aristotle's theory of gravity stated that objects fall at a speed relative to their mass.
110. Peter Graneau PhD, "Is Dead Matter Aware of Its Environment" *Frontier Perspectives* magazine, Vol 7 No. 1, p. 51.
111. Http://shadaether.tripod.com/shad.htm.
112. Milo Wolff PhD *Exploring the Physics of the Unknown Universe: An Adventurer's Guide*, Technotran Press, 1600 Nelson, Manhattan Beach California, 1990 p. 178.
113. J. A. Wheeler and C. Misner, *Geometrodynamics*, Academic Press, New York, 1962.
114. Dan Davidson's book is *Shape Power: How Form Converts Universal Aether into Electromagnetic and Gravitic Forces*. Now he has a new booklet, *Shape Power and Universal Resonance*.
115. Sacred geometry is a concept popular in metaphysical circles.
116. "Scale Unification - A Universal Scaling Law for Organized Matter" *Proceedings of the Unified Theories Conference in Budapest, Hungary, 2006.*
117. Michelson and Morley's interferometer experiment.
118. *Michelson and The Speed of Light*, Bernard Jaffe, 1960.
119. *Ether-Technology*, Cadake Industries, Georgia, 1977, p 65.
120. Correspondence between Maurice Solovine and Einstein on March 14, 1949, quoted in B. Hoffman *Albert Einstein: Creator and Rebel* 1972, p. 328.
121. "Dayton Miller's Ether-Drift Experiments: A Fresh Look" by James DeMeo, PhD http://www.orgonelab.org/miller.htm.
122. April 22, 2002 phone conversation between Gene Mallove and Jeane Manning.
123. Www.aetherometry.com.

124. Valone, Thomas, Zero Point Energy: The Fuel of the Future. Integrity Research Institute, Maryland, 2007.
125. Http://www.interstellartechcorp.com.
126. Brian Handwerk, "Antigravity Machine Patent Draws Physicists' Ire" *National Geographic News*, November 11, 2005 http://news.nationalgeographic.com/news/2005/11/1111_051111_junk_patent.html.
127. Eugene Mallove "The Strange Birth of the Water Fuel Age: The Cold Fusion 'Miracle' was No Mistake" February 2000 on request from the White House.
128. Stringham and business partners formed a company, EQuest, which eventually was dissolved. Later he and Dick Raymond formed First Gate Energies. Its mailing address: P.O. Box 1230, Kilauea, HI 96754 USA.
129. Technical note: In his "Cavitation and Fusion" paper written for the tenth International Conference on Cold Fusion, Stringham said sonofusion is an "emergent tangent technology to sonoluminescence technology which we use to give us an environmental parameter probe into the bubble contents at the moment of its highest energy density."
130. At this time the acoustic power oscillator of Stringham's experimental device relies on 60-cycle electricity from the wall, but he expects the device could be liberated from any input power, with the exception of a battery to start the process.
131. Stringham figures that the tiny devices could be etched on a durable material, silicon dioxide, and stacked in perhaps an eight-inch cube to create significantly large amounts of heat output.
132. Www.lenr-canr.org.
133. The New Energy Times *Special Report on Bubble Fusion/Sonofusion* July 10, 2007.
134. Stringham: "The reaction product of this fuel is helium with no long range radiation by-products as measured in the Los Alamos National Laboratory and in my twenty years of experience working with this cavitation process. Like the puzzle of sonoluminescence, lack of radiation products in the sonofusion phenomenon is explained by several published theories and is still controversial."
135. The actual range of temperatures inside experimental cold fusion apparatus varies.
136. Like charges repel, and the repulsion between protons is called the Coulomb barrier.
137. Ed Wall, "Device and Process Testing Update", *Infinite Energy* issue 29, 1999 p. 52.
138. Journalist's name not available.
139. Grants from the Army's research office and from New York Community Trust.
140. John Dash interview with the author, January 2000.
141. Www.infinite-energy.com.
142. August 25–29, 2003 at MIT's Department of Electrical Engineering.
143. A laboratory owned by Energetics Technologies LLC of New Jersey.
144. Laboratories listed by Steven B. Krivit in article in April 19, 2008, Great Falls Tribune, www.greatfallstribune.com.
145. Stringham, Roger, "Low Mass 1.6MHz Sonofusion Reactor", Proceedings of International Conference on Cold Fusion 11, 2005.
146. Technical note: Rauen's heat engine concept is based upon the experimental evidence of the Proell effect, a macroscopic Maxwell's Demon.
147. Team led by Stanislaw Spzak and Pamela Mosier-Boss at SPAWAR.

148. The co-deposited film was their cathode.
149. Called the CR-39 detector.
150. Swartz, M, "Improved Electrolytic Reactor Performance Using p-Notch System Operation and Gold Anodes, *Transactions of the American Nuclear Association*, Nashville, Tenn Meeting.
151. Krivit, Steven B. "Cold fusion is neither dead nor merely a wishful fantasy", Op-Ed page April 19, 2008, *Great Falls Tribune*, www.greatfallstribune.com.
152. Jed Rothwell "Cold Fusion and the Future", *Infinite Energy* magazine # 12, Jan.–Feb. 1997 p 11.
153. Newmyer, Jacqueline, Harvard Crimson, 5/17/2000.
154. Stolper, Thomas, Genius Inventor, self-published 338 page book, 2006, page 303.
155. Thomas S. Kuhn, *The Structure of Scientific Revolutions*, University of Chicago Press, 1970, p. 90.
156. Telephone interview with Jeane Manning, Feb 10, 2000.
157. BlackLight Power Inc. press release May 28, 2008.
158. According to Mills' 178 page technical presentation in 2008, at www.blacklight power.com, BLP Energy Technology's installed cost is $250/Kw vs. $700/Kw for an industrial gas turbine. Nuclear is $2,200/Kw.
159. Randell L. Mills, *The Grand Unified Theory of Classical Quantum Mechanics*, Science Press, Ephrata PA 1996.
160. Potassium.
161. Mills' 178 page technical presentation 2008, www.blacklightpower.com.
162. BlackLight Power Inc press release May 28, 2008.
163. BlackLight Power Inc. press release May 28, 2008.
164. Stolper, Thomas, *Genius Inventor*, self-published 338 page book, 2006.
165. Manning's interview with Mills.
166. Stolper, Thomas, *Genius Inventor*, self-published 338 page book, 2006, page 86.
167. Stolper p. 87.
168. Erik Baard, "The Quantum Leap", *The Village Voice*, published online Dec. 22–28, 1999.
169. Interview with Manning.
170. "Randell Mills—New Energy and the Cosmic Hydrino Sea", interview by Art Rosenblum, *Infinite Energy* Dec. 1997–January 1998 p 32.
171. Free-electron lasers.
172. Professor Herman Haus of the Dept of Electrical Engineering and Computer Science, MIT.
173. Manning interview with Mills.
174. Stolper p. 90.
175. Which included a universal drug-delivery molecule and a body scanner.
176. Manning interview with author Feb. 10, 2000.
177. Robert Schaubach.
178. 1992 Intersociety Energy Conversion Engineering Conference.
179. "Hydrogen is Potential New Energy Source", Reuters Information Service, April 1, 1997.
180. "Hydrogen is Potential New Energy Source", Reuters Information Service, April 1, 1997.

181. Stolper, p. 306.
182. New Journal of Physics.
183. Referring to the folk tale in which a child points out a truth which adults ignored, either from fear of their ruler or through distrust of their own perceptions.
184. In Greifswald, Germany, Conrads had become the director of the Institute for Low Temperature Plasma Physics at Ernst Moritz Arndt University.
185. Manning's interview with Mills.
186. *Energy Invention Suppression Cases* compiled by Gary Vesperman, engineer, self-published 2007 on Internet.
187. Eugene Mallove, introducing Peter Graneau's speech at Cold Fusion Day at MIT, 1995.
188. Robey, James A., *Water Car: How To Turn Water Into Hydrogen Fuel*, Kentucky Water Fuel Museum 2006, p. 17.
189. Kanarev, Ph. M. "Low Current Process of Water Electrolysis, 4/8/2003 http: //guns.connecgt.fi/innoplaza/story/Kanarev/lowcurrent.
190. Http://www.xogentechnologies.ca/.
191. H.G. H. Hermans, *Memories of a Maverick*, Pi Publications, Maassluis, The Netherlands, 1998 p. 11.
192. Puharich, US Patent 4,394,230 "Method and Apparatus for Splitting Water Molecules".
193. Andrew Michrowski "Puharich water splitting system for cars" PACE Newsletter Vol. 8 No. 2, p. 14.
194. Elizabeth Rauscher, PhD "Enhancing Car's Electric Drive" PACE Newsletter Vol. 8 No. 2, p. 14.
195. Elizabeth Rauscher, PhD "Keeper of the Fires of Learning" PACE Newsletter Vol. 8 No. 2, p. 9.
196. Swiss Assoc. for Free Energy, Einsiedeln 1988.
197. 1985 videotape, reportedly held by London University.
198. Sept. 11, 1994 letter from Stan Meyer to author, enclosing copies from books: by Colin A. Ronan, *Science Explained—The World Of Science in Everyday Life*: Henry Holt NY 1993 p 112-116; and by Harold Nathan, Chemistry, Cliffs Notes USA1993 p 51-2; 97-8.
199. Stanley A. Meyer "Water Fuel Cell: International Independent Test-Evaluation Report 1993, p 13.
200. Hyperdrive was to apply energy made available from high-voltage pulsing of the zero-point energy of space, directly into generating a water jet. Hyperdrive would not have required an engine and would have had no moving parts. It would have been significant in propelling ships, since steering would have come from feeding the jet through maneuvering nozzles.
201. Jeane Manning's notes from telephone conversation with Gene Mallove, October 23, 1995 concerning Mallove's visit two years earlier.
202. James A. Robey *Water Car: How to Turn Water Into Hydrogen Fuel* p. 79.
203. Http://www.gasandoil.com/goc/features/fex51230.htm.
204. Ph. M. Kanarev, G.P. Perekoty, D.A. Bebko, A.A. Chernyavsky "Water Electric Generator of Heat".
205. Http://www.hydrodynamics.com/index.htm.
206. David Wallman U.S. Patent No. 5,417,817.

207. Http://www.clean-air.org/.
208. Www.magnegas.com.
209. Monatomic oxygen and monatomic hydrogen.
210. US Patent 3,310,483 Multicell Oxyhydrogen Generator.
211. Www.blogtalkradio.com/kywaterfuelmuseum.
212. Www.callowayengines.com.
213. Magnets in Your Future Magazine 1988.
214. Harold Puthoff PhD of Austin TX is a leading author of such peer reviewed papers.
215. Haramein, Nassim What Is the Origin of Spin http://theresonanceproject.org.
216. Www.mullerpower.com.
217. The late Dr. Carl Reich.
218. Http://earlyradiohistory.us/1920alt.htm.
219. Space Technology and Applications International Forum conference, University of New Mexico, February 2006.
220. By the Hants and Berks Gazette 1973–1975.
221. Thomas, John A., Antigravity: The Dream Made Reality, Direct International Science Consortium, 1993.
222. Dr. Anatoli Rykov, chief of the Seismometyr and Engineering Seismology Lab of Moscow "There are No Secrets in Inertialess Motion and Antigravitation", *New Energy Technologies*, January 2002, page 38.
223. V.V. Roschin and S.M. Godin, *Technical Physics Letters*, Vol. 26, No. 12, 2000 "An Experimental Investigation of the Physical Effects In a Dynamic Magnetic System".
224. http://news.independent.co.uk/uk/environment/story.jsp?story=523735.
225. www.free-energy.cc.
226. Ken Rauen September 21, 2004 email reply to Jeane Manning.
227. www.NewEnergyCongress.org.
228. http://environmentalfame.com/.
229. Majumdar, Yang, Hochbaum, Chen, Diaz Delgado, Liang, Garnett, Najarian "Enhanced Thermoelectric Performance of Rough Silicon Nanowires." Nature January 10, 2008.
230. Www.lbl.gov/Science-Articles/Archive/MSD-silicon-nanowires.html.
231. US Patent 6,396,191, 2002.
232. Thomas Valone www.integrity-research.org/gave us this information about Hagelstein's diode.
233. The inventor of the novel fuel and his company do not want publicity at this writing, but did agree on this general announcement.
234. Moray King, The Energy Machine of T. Henry Moray and Quest for Zero-Point Energy.
235. Www.aetherometry.com The Correas investigated the "pulsed abnormal glow discharge" that occurs when electrons come off of a cold cathode in a vacuum.
236. Http://www.rexresearch.com/correa/correa.htm.
237. Gene Mallove in conversation with Jeane Manning.
238. Http:// www.svn.net/krscfs/.
239. Www.magneticpowerinc.com.
240. Www.startech.net.

241. Compressed air vehicle developed by Guy de Negre of France.
242. Senator Maria Cantwell on the floor of the US Senate, 15 May 2008, immediately after the Senate passed a resolution to overturn the FCC's Dec. 07 decision to deregulate media cross-ownership.
243. Energy From The Vacuum DVD, Part 2.
244. The Monopole Motor group.
245. A flywheel is not a factor in the SG motor but was on a separate design built by Jim Watson around 1984.
246. Co-author Jeane Manning empathizes with those researchers who have been used by the media. A CBC television news program in the late 1990s asked to send a film crew to Manning's home. Before the interview the crew blacked out the large windows. Her freshly-painted office was suddenly in darkness. They admitted to intending to create an X-Files atmosphere, which they did. The effect conveyed "conspiracy theorist".
247. Westdyk, Karin, "U.S. Patent Granted To West Milford Resident for Unique Hydrogen Generation System" The Messenger: Environmental Health Journal, March/April 1922, 13–15.
248. Lambertson, Wingate, "Do We Need Nature?" Explore magazine Volume 12, Number 4, 2003.
249. Andrew Michrowski email to author November 9, 2003.
250. Tosches, Rich, *Gazette Telegraph*, Colorado Springs, July 25, 1993.
251. Milton, Richard, *Alternative Science: Challenging the Myths of the Scientific Establishment* Park Street Press, Vermont, 1996, preface.
252. Facts on who controls the media are found at www.nowfoundation.org/communications/tv/mediacontrol.html.
253. Www.mediagrassroots.net/.
254. Http://www.reclaimthemedia.org.
255. Tom Bearden interview with coauthor Jeane Manning, 2004.
256. Www.focusfusion.org.
257. IEEE-APS International Conference on Plasma Science in Banff, Alberta, Canada.
258. Tim Ventura interview with inventor ww.americanantigravity.com/documents/Quasiturbine-Interview.pdf.
259. Reports say the four-blade rotary turbine spins without vibration or dead zones yet with strong torque even at slower speeds. It is compatible with multiple fuels including hydrogen, with no need for fuel additives.
260. Photo-detonation is an optimal combustion mode of the future that the piston engine can't support. See www.quasiturbine.com for more.
261. His company Quasiturbine Agence Inc attended an international conference for automotive engineers in Wisconsin and a major auto show in Munich in 1999. The latter invitation arrived as a result of a cover story about in a major European automotive magazine.
262. The following anecdotes are adapted from the email correspondence between Jeane Manning and Saint-Hilaire December 31 2007–January 4, 2008.
263. One such group is the Ontario Quasiturbine Application Promotion Association http://www.promci.qc.ca/pureinvention/oqapa/.
264. John Hutchison conversation with Jeane Manning.
265. Moray King speaking at TeslaTech in Salt Lake City, July 2004.

266. Stephan Herrera "Professor Smalley's Latest Big Idea: Nano-energy Will Save The Earth", Small Times Aug. 28, 2002.
267. In the year 2000.
268. Shawnee Baughman interviewed by Jeane Manning.
269. Http://www.icehouse.net/john34/.
270. John Bedini and Tom Bearden, Free Energy Generation Cheniere Press 2006.
271. Www.energenx.com.
272. Described as a unusual "Kromrey gravity-field generator" powered by a battery bank that virtually never needed charging from an outside source; the system kept its batteries charged. Bedini said it tapped zero point energy or the "vacuum" by using stressed, pulsed scalar waves out of phase with each other. The concept was not found in textbooks but was perfectly natural and it works, he said.
273. Sisson is on NEM Board of Directors.
274. Couch, Michael, "Bob Boyce's Electrolyzer Plans Available to All" Pure Energy Systems News January 8, 2007.
275. Http://www.shinyeonenergy.com/.
276. Andrew Mount on Television show, *The Next Step*, show N. 53 with Daniel Kottke, topic The New Physics.
277. Elizabeth Rauscher phone conversation with Jeane Manning, around 1995. Dr. Rauscher did not give details of the meeting's purpose or specific topics.
278. Speaking on a panel at the New Energy Movement conference, 2004.
279. Integrity Research Institute, Washington DC.
280. Ted Loder PhD on William Alek radio show talking about the Orion Project May 10, 2008. http://www.theorionproject.org/.
281. Www.newenergymovement.org and www.newenergymovement.ca.
282. Www.NewEnergyCongress.org.
283. Http://FreeEnergyNews.com June 19, 2008.
284. Shinyeon Energy Research Center.
285. Jean Houston for her "Living the New Story: A Renaissance of Self and Society" weekend workshop.
286. Bruce Sterling, in an interview in the WELL's Inkwell.vue forums, January 2000.
287. From Bruce Sterling speech delivered 9/17/98 as a part of a ZDNet/Arthur Andersen "Next 20 Years" event.
288. Chip and Dan Heath *Made to Stick* Random House, 2007.
289. Resources include www.gan-net.net Global Action Network.
290. Www.ahealedplanet.net.
291. NEM conference in Portland, Oregon 2004.
292. Excerpted from the lyrics of "I Choose Love" by Shawn Gallaway, with permission.
293. Http://www.aquapol.co.uk/.
294. Www.aquapol.at.

RESOURCES

A sampling of informative websites:

aetherometry.com Papers written by Paulo and Alexandra Correa, on massfree energy, Tesla, Wilhelm Reich's work.

americanantigravity.com Interviews with scientists, well written.

blacklightpower.com About a new field of novel hydrogen chemistry and energy production.

cheniere.org Thomas Bearden's website with a wealth of essays on how the energetic "vacuum" of space could be tapped.

eagle-research.com George Wiseman's hydroxy research and fuel saving information.

energyscience.co.uk Authoritative and very educational site by Harold Aspden PhD.

evert.de Site from Germany in English, visuals.

faraday.ru Alexander Frolov's lab and New Energy magazine, Russia.

free-energy.ws/—Peter Lindemann's site; technologies and new energy history. He has replicated many effects.

FreeEnergyNews.com Sterling Allen's daily updates on clean energy developments.

halexandria.org/ Synthesis of new physics, sacred geometry, ancient and modern history.

hasslberger.com/ Commentaries from Italy in English, with new energy and philosophy.

icehouse.net/john 34 John Bedini's website with various of his proof-of-concept test model machines and commentaries.

infinite-energy.com Infinite Energy magazine.

integrity-research.org Integrity Research Institute and Thomas Valone's e-newsletter.

interstellartechcorp.com Fabrizio Pinto PhD and the physics of zero point energy.

jnaudin.free.fr Jean-Louis Naudin's virtual laboratory.

KeelyNet Free Energy (http://escribe.com/science/keelynet/) Information on site dates back to the earliest computer bulletin boards and is updated regularly.

lenr-canr.org Technical papers on cold fusion.

magneticpowerinc.com Mark Goldes' company site, essays on harnessing magnetic power.

MullerPower.com Information about magnetic knowledge of the late Bill Muller.

Newenergymovement.ca Posts news of upcoming new energy events in Canada and links to videos and articles.

Newenergymovement.org Contains essays on the implications of new energy and a full copy of the draft legislation for an Energy Innovation Act.

newenergytimes.com News site on the LENR, cold fusion field.

padrak.com/ine Papers from the now defunct New Energy News and Institute for New Energy database.

peswiki.com New Energy Congress; see the Top 100 listing, updated regularly.
rexresearch.com A collection of hard to find papers.
SeasPower.com Space Energy Access System—Dr. Steven Greer's group.
SVPvril.com Sympathetic Vibratory Physics; the science of sound and vibration.
TheResonanceProject.org Nassim Haramein and the Resonance Project.
Teslatech.info Steve Elswick's website for Tesla Tech conferences and books.
zpenergy.com An engineer's focused news site for zero point energy developments. Watch this site for news of an Xtreme Science Foundation prize for inventors. This is a very worthy project proposal and at this time needs skilled volunteers to set up the foundation and its transparent bookkeeping and needs celebrities to publicize the opportunity for the public to become involved in the project. Contact zpenergy.com

GLOSSARY

Aether. The background substance of the universe, now thought to be in a constant spiralling motion, that of a vortex. It is the basic substance out of which the universe is made, and it gives rise to space energy. Currently an unorthadox concept.

Alternating current (AC). Electricity that flows back and forth in a regular rhythm. In the United States and Canada, standard household current changes direction sixty times a second.

Antigravity. A force that opposes or cancels gravity, which is the force that pulls or pushes all objects on the earth's surface towards the planet's center. Under certain conditions, space energy can produce an antigravity effect.

Casimir effect. The tendency for two perfectly smooth metal surfaces placed very near each other to come closer together. It is thought that space energy causes this effect.

Cavitation. The formation of cavities or bubbles in liquids, and the collapse of those bubbles. Shock waves are created when the bubbles that form in a low-pressure section of a liquid-carrying pipe collapse upon being carried to a high-pressure section. Cavitation is also called water hammer.

Charge cluster. A ring-shaped structure made of tightly packed electrons.

Chemical energy. Energy produced by burning, such as that produced by oil or coal.

Closed system. A system in which a finite amount of energy is available. An internal combustion engine can be said to operate in a closed system.

Cold fusion. The joining together of atomic nuclei under room temperature conditions in order to release energy.

Direct current (DC). Electricity that flows in one direction.

Dynamic electricity. Electricity in motion, such as the flow of current through a wire.

Earth resonance. The sending of electric pulses through the earth at the same rate at which the earth itself vibrates in order to build up large waves of energy. Such energy could be picked up at a distance by an antenna.

Electrolysis. The breakdown of water into its component parts, oxygen and hydrogen, by passing an electric current through it.

Electrolytic cell. A cell that contains an electricity-conducting liquid in which metal wires or plates are suspended.

Electromagnetics. The physics of electricity and magnetism.

Entropy. The idea that matter and energy are always becoming more and more disorganized. It is the opposite of negentropy.

Fission. The splitting of an atom's nucleus in order to release energy.

Fossil fuel. A controversial concept recently as some scientists argue for other processes creating these fuels, but geologic stores of carbon fuels are long thought to be the fuels produced by the decay of prehistoric plants and animals deep underground. Oil, coal, and natural gas are all considered fossil fuels.

Free energy. See space energy.

Fuel cell. A cell in which a gas, such as hydrogen, is used to create electricity.

Generator. A device that converts mechanical energy into electricity.

Heat pump. A device that heats a structure by drawing heat from the surrounding soil, air, or water.

Heat technology. Devices that derive energy from a difference in temperature, for example, by changing a liquid into a gas. Some new energy heat technologies use liquids that change into gases at low temperatures.

Hot fusion. The joining together of atomic nuclei under conditions of high heat and pressure in order to release energy.

Hydrogen. The lightest known element, consisting of one proton and one electron. Its abundance and wide distribution over the earth's surface means that hydrogen can be used to provide inexpensive, decentralized energy.

Hydropower. The use of moving water to generate electricity. Standard hydropower uses environmentally destructive dams, but new energy water power approaches use devices that do not harm the environment.

Implosion generator. A generator that uses an inward-spiralling vortex movement to produce power.

Induction. The electrification of a wire that occurs when the wire is placed near a moving magnetic field.

Internal combustion engine. Vehicle engines in which refined carbon fuel, such as gasoline, is burned. The energy released by this burning is transformed into a rotary motion that propels the vehicle's wheels.

Magnet device. A device that uses magnets to turn space energy into electricity.

Magnetic drag. A problem in standard generators in which residual magnetism slows down the rotor, which is the part that either moves the magnets past the coils of wire or the wire coils past the magnets—see induction. This drag reduces the generator's output.

Metal hydride. A combination of metals that allows for the safe storage of hydrogen. Under proper conditions, the hydride soaks up hydrogen and holds it until the hydrogen is released on demand.

Motor. A device that converts electricity or chemical energy into mechanical energy.

Negentropy. The idea that matter and energy can organize themselves. It is the opposite of entropy.

New energy. Energy that comes from nonconventional sources, preferably that which can be produced on a decentralized basis. Cold fusion, heat technology, hydrogen technology, low-impact hydropower, and space energy are all examples of new energy.

Nuclear energy. Energy produced by either the breaking apart or bringing together of atomic nuclei. Also see cold fusion, fission, and hot fusion.

Open system. A system in which an infinite amount of energy is available. A space energy device can be said to operate in an open system.

Overunity. A condition in which there is more energy going out of a device than coming in.

Perpetual-motion machine. A device that, once set in motion, continues to operate without an outside source of energy within a closed system. Such a device is by definition impossible to build.

VTA can be resurrected so that the world will realize what a pioneer Floyd Sweet was.

Confusion and secrecy

Despite Bearden's urging, Sweet never had the VTA certified by independent testing. "He feared that his life would be snuffed out immediately if he even attempted such a thing," Bearden said. Sweet also frustrated his fellow researchers by keeping secret his most important process—how he conditioned the magnets that are at the heart of the VTA. Did he pump the magnets with powerful electromagnetic pulses to shake up their internal structure? He refused to give details, and said it wasn't likely that other researchers would learn his secrets.

Sweet not only feared for his life, but once said he feared that if he described how he made his device work, unscrupulous people would build models without giving him his due. He was also concerned about what would happen if the VTA was widely sold everywhere at once, replacing many other electric devices. "If it all came out at once, the stock market would collapse," he said. "The government doesn't want it."

A combination of such attitudes is sometimes called "inventors' syndrome" by business people and others who have tried to help move energy inventions from the proof-of-concept stage to the marketplace where you could buy an energy converter for your home. It is frustrating for would-be helpers when inventors retreat into fearfulness, hide crucial details of their inventions and die with their secrets. To be fair to Sweet, we would point out that he is not the only inventor who has been uncomfortable in disclosing key aspects of his work.

Other researchers and the VTA

Other inventors have tried to carry on Sweet's work, especially in the months before and after his death. In Somerset, England, electronics expert Michael Watson built a replica of Sweet's VTA, but claimed no success in the experiment. Despite that, he said, "In my opinion the inventor of the VTA, Floyd Sweet, has made a scientific discovery of the greatest importance."

Watson indicated that attempts to reproduce Sweet's results may run into problems because the type of magnets Sweet used are no longer available. But he said, "The important point about the VTA is that a form of magnetic instability exists that can act as a significant energy source."

Quantum mechanics. The branch of science that deals with protons, electrons, and other basic particles of matter.

Self-oscillation. The continuous shaking or vibration of a magnetic field.

Solid-state devices. Devices that contain no moving parts.

Sonoluminescence. The light that is given off when ultrasound waves are pumped into tiny bubbles in a liquid, and the bubbles collapse violently.

Space energy. Energy that consists of electrical fluctuations in the aether. It is present everywhere in the universe, including the earth, but can only be put to use through the use of specially designed devices.

Static electricity. Electricity at rest, such as the electric charge that builds up on a plastic comb.

Superconductivity. The state of a wire when it suddenly loses resistance, the force that keeps current from passing through the wire. Superconductivity normally occurs only at very low temperatures, but it could help create a practical source of electrical power if it could be made to occur at room temperature.

Turbine. A machine that uses a stream of either gas or liquid to turn a shaft, such as the rotor of a generator.

Vortex. A three-dimensional spiral that creates a funnel of energy, such as a tornado. Motion along such a spiral can be outward, in which energy is dissipated, or inward, in which energy is created.

Zero-point energy. See space energy.

ABOUT THE AUTHORS

Joel Garbon received a BS in Applied Science from Miami University, Oxford, Ohio. He has over two decades of experience in product development and as a technical consultant to the paper, chemical, building products, and water treatment industries. He has consulted with dozens of large and small companies in North and South America and is a highly regarded speaker and educator within these industries, regularly instructing classes of engineers. He developed several commercially successful chemical formulations for use in paper and building products applications and is co-inventor of a water treatment filtration technology.

Joel currently serves as President of The New Energy Movement (www.NewEnergyMovement.org) and is a founding member of New Energy Congress (www.NewEnergyCongress.org), which maintains a continuously updated and publicly viewable database of the "Top 100" most promising breakthrough energy technologies. Joel has relationships with many organizations which promote responsible stewardship of earth's resources and which support progressive sustainable technologies for advancement of human civilization. He has been a featured speaker on numerous television and radio programs and at conferences and civic gatherings, making the appeal for a concurrent evolution in human consciousness and energy technology to address our pressing global challenges.

He is the author of an historic legislative draft titled "Energy Innovation Act of 2007" which is intended for introduction as a Congressional bill and to serve as template for new energy legislation on the state, local, and international levels. The legislation's provisions call for urgent and serious federal support for research and development of breakthrough energy technologies.

Jeane Manning is a long-time researcher of quantum-leap energy systems that could replace oil, and their implications for humankind. She authored The Coming Energy Revolution (Avery Publishing Group, New York, 1996, now out of print); Energie (Omega Verlag, Gemany, 2000); and coauthored other nonfiction books including Angels Don't Play This haarp, co-authored with Dr. Nick Begich. These books are published in six languages and she has been an invited speaker at several energy conferences in Europe.

Jeane began to chronicle the new energy scene in the 1980s. During the years while she was staff writer for a regional lifestyle magazine, edited a small town newspaper and held other employment, she used her vacation time to interview scientists in North America and Europe. She has served on boards of directors for a new energy institute that was based in Utah, the Institute for New Energy; ocean energy company Blue Energy Canada; geothermal company Essential Innovations; and the New Energy Movement based in Portland, Oregon.

In January of 2007 a group of men and women gathered at Jeane's apartment in Vancouver, Canada, to form New Energy Movement Canada. These grassroots efforts are intended to help bring about a world of clean air, vitalized waters and healthy soils. Jeane's three grown children and their young families inspire her efforts.

INDEX

A

Adams, Robert, 92, 104
agri-business, 38
Alexanderson alternator, 172
Allen, Sterling, 177, 229
Alternative Science: Challenging the Myths of the Scientific Establishment, Richard Milton, 199
American Hydrogen Association, 167
Aquarius Now, Marilyn Ferguson, 42
Arata, Yoshiaki, cold fusion, 145
Aspden, Harold, 102, 120
  ether, 130
asthma, 29
attitude. *see also* grassroots groups
  biased, 194
  changes in, 47–49, 228

B

barium titanate, 184
Baughman, Shawnee, science fair project, 214
Baumgartner, William
  implosion technology, 72
  Wavy Plate, 127
Bearden, Thomas, 95, 119, 201
  electromagnetic generator, 184
  extracting energy from the "vacuum", 124
  Floyd Sweet's Secrets, 101
  SG motor, 215
  Vacuum Triode Amplifier, 97
Bedini, John, 95, 204–205, 213
  School Girl Motor system and the media, 189–192
  SG motor, 214–216
biodiesel, 38, 39
biofuels, 38–40
  alternate sources of, 39, 40
BlackLight Power Inc., 149
  BlackLight process, 156–157
Blue Energy, tidal power, 111, 114
Boadella, David, 82
Bockris, John, 28
boron, 202
Boyce, Bob, open sourcing, 225
brain-trust, worldwide ripple effect, 15, 221
Brown, Yull, Brown's gas, 167
Burger, Martin, 43
Bushnell, Dennis
  atmosphere-pollution crisis, 231
  energy-related problems, 39–40

C

Calloway, Robert H., 169
CANR (chemically-assisted nuclear reactions), 140
Cantwell, Senator Maria, 189
capitalists, venture, 217, 219
carbon dioxide, 29–30, 38
  sources of, 37
carbon neutral, 38
cavitation, 135–137, 185
China, xi
Chukanov, Kiril, ball lightning, 182–183
citizens, groups, 199, x. *see also* grassroots groups
Coler, Hans, magnetic-energy generator, 102
Coming Energy Revolution, The, Jeane Manning, vii
Comings, Mark, 49–50
Conrads, Johannes, 156
converters, ix, x. *see also* energy, new technology
Cook, Nick, *Jane's Defense Weekly,* 15

corn-to-ethanol, and agri-business, 38
Correa, Paulo and Alexandra, 91, 131 181–182 .
Couch, Michael, 213
  Bob Boyce and monatomic hydrogen, 224–225

D

dam, hydro, 8
Dash, John, 134
  cold fusion cells, 142, 143–144
Davidson, Dan A.
  Shape Power, 50
  shapes resonate, 127
deGaulle, Charles, president of France, 110
DelloRusso, Barbara, *Infinite Energy*, 108
DeMeo, James, Dayton Miller, 131
DePalma, Bruce, 92
deuterium, 137–138, 139, 143, 145
Dingel, Daniel, water-car, 164–165

E

Earth, magnetism of, 246
Earthsave, 207
ecosystems. *see* environment
Edison, Thomas, 58
Einstein, Albert
  ether, 130
  etherless universe, 117
Electric Council of New England, The, 108
Electric Ride, 168
electricity, free, xi. *see also* energy, new technology
electromagnet, 170. *see also* energy, new technology
Eneco Inc., 179
energy
  alternative. *see* energy, new technology
  corporate vested interests, 38–40, 219, 230
  invisible, 24–25
  law of conservation of, viii
  literacy, 4, x
  new technology, 24–25, 50
    benefits of, 46–48
    changes in attitude, 49, 194, 228
    financing costs of, 217, 219–220
    government awareness of, 52
    obstruction or opposition to, 13, 60, 63–64, 68, 90, 99–100, 104, 108, 112, 120, 132, 138, 155–156, 189–200, 208–209, 215–216, 219–220, 223
    patents for, 218, 220–223
    research, 1, 16, 18, 34–37, 43–48, 123, 159–168, 177–186, 234
  problems related to, 39–40
  quantum-leap revolution, 48–49, 221
  renewable, 1
Energy Autonomy, Hermann Scheer, 33–35
Energy Innovation Act, 19–20
  draft legislation, 237–242
entropy, 98
enviro-activists, x. *see also* grassroots groups
environment
  greenhouse gases, 37
  a higher civilization, 43–45
  polluting toxins, 231, ix
  remediate damage, 39–40, 221, 230–232
  tipping point, 38
  World Environment Day, 177
Environmental Hall of Fame, 177
Era
  Hydrogen, 160
  New Energy, 14, 46–48
ethanol, 38
Ether Technology, Rolf Schaffranke, 11
Exotic Vacuum Objects, 183

F

Faraday, Michael, Faraday's Law, 7, 160
Felenstein, Lee. *see* Gunderson, Graham
Ferguson, Marilyn, *Aquarius Now*, 42
fission, 138
Fleischmann, Martin and Stanley Pons, cold fusion, 141–142
Flynn, Joseph, 172–173
Flynn Research Inc., 172

footprint, ecological, 39–40, 207, x
Francoeur, Alan, interference disc generator, 102, 171–172
Frazier, Wade, wise choices of energy, 234
Free Energy Updates , 230
fuels, carbon. *see* fuels, fossil
fuels, fossil, 25. *see also* energy, corporate vested interests
  cost of, 39–40, 46
  and greenhouse gases, 37
fusion, 138, 139, 141–146, 182

G

Gallaway, Shawn, *I Choose Love*, 234
Garbon, Joel, 13, 28, vii
  Energy Innovation Act, 19–20, 229, 237–242
  and new energy technologies, 35–36
Gary, Wesley W., fuel-free magnetic motor, 101–102
gas, hydrogen, prototype systems for, 160–168
gases, greenhouse
  and the Earth's tipping point, 38
  sources of, 37
Generating Electricity by Nondestructive Interference of Energy (GENIE), 32. *see also* Gunderson, Graham
generators, 185, ix
GENIE (Generating Electricity by Nondestructive Interference of Energy), 32
Genius Inventor, Thomas E. Stolper, 148
Ghandhi, Mahatma, 226
Gilbert, Sir William, 11
Goddard, R.H., 11
Godin, Sergi and Vladimir Roschin, 174–175
Gore, Al, 1
Graneau, Peter, 121
  exploding water, 184
grassroots groups
  opportunities for activism, 39–40, 229, 232–234
  stewards of the planet, 35, 43–45, 47–49, 230–231
Gray, Edwin Vincent, cold electrical energy, 103
Green Era, 1
Green Power Inc., 185
Greer, Steven, 42, 228
Griffin, Sir Anthony, *Water as Fuel*, 164
Groth, Phil, MadeByKids Foundation, 216
Gunderson, Graham
  early, 30–31
  GENIE, 32–33
  hampered by lack of funding, 208
  oscillating magnetic fields, 101

H

Hagelstein, Peter, 179
Haisch, Bernard, sea of light, 132
Haramein, Nassim, geometry of space, 128–129, 170
Hawken, Paul, 18
Heisenberg, Werner, 169
helium, 139, 146
Hilscher, Gottfried, 3
Hiroshima, 12
Houston, Jean, stewards of the planet, 230–231
Hutchison, John, crystal converter, 184, 205
hydrogen, 160, 167, 184, 192–193, 224–225

I

I Choose Love, Shawn Gallaway, 234
India, xi
Infinia Corporation, concentrated solar energy, 185
Inomato, Shiuji, 245, 246
Intention Experiment, The, human thought and intention, 43
Interference Disc Generator, 171–172
International Association for New Science, 197
International Panel of Climate Change (IPCC), and greenhouse gases, 37
inventions. *see* energy, new technology
inventors, as consultants, 223–224
IPCC (International Panel of Climate Change), and greenhouse gases, 37

J

Jane's Defense Weekly, 15
Japan, xi
Jenkins, Bill, free energy generator, 215
Johnson, Howard R., asymmetric magnetic systems, 93, 104–105
Josephson, Brian, qualities of the invisible energy source, 129

K

Kanarev, Philip M., plasma electrolysis, 165–166
Kanzius, John, radio waves, 166–167
Kaplan, Steve, the media and debunkers, 195, 199, 229
Keely, John, vibration, 83–86
Kennedy, John F. president of United States, Passamaquody Tidal Power Project, 107–108
King, Moray B., 3, 65, 70–71, 181, 208–209
Kuhn, Thomas S., *The Structure of Scientific Revolutions*, 148

L

Lambertson, Wingate, the media and universal dark energy, 195
Laszlo, Ervin, standing waves, 124
Leahy, Adm.William D., *The Truman Memoirs*, 12
LENR (low energy nuclear reactions), 140, 145
Lerner, Eric
  hampered by corporate interests, 208
  opposition to plasma science, 202
  plasma focus fusion, 182
Lindemann, Peter, 92, 104, 176, 204–205, 227–228
Loder, Ted C., III, 229
Loncar, Zlatko "Shad"
  converted energy, 124
  free energy, 122
  vortex system, 166
Ludwig, Thorsten, dehydration units, 247

M

Made to Stick: Why Some Ideas Survive and Others Die, 232
MadeByKids Foundation, 216
MagneGas, 167
Magnetic Power Inc.. *see* Gunderson, Graham
magnets, power of, 170–175
Mallove, Eugene, 91, 134, 159
  cold fusion, 144
  Infinite Energy, 144
  The New Heretic, 197–199
  Paulo and Alexandra Correa, 182
  Stanley Meyer, 164
  zero-point energy, 131
Manning, Jeane
  Bill Muller, 8
  Blue Energy, 114
  fiery crisis, 231
  hampered by corporate interests, 216
  John Bedini, 192
  The Coming Energy Revolution, vii
  The New Energy Movement, 4
  opposition to technical solutions, 206–207
  Passamaquody Tidal Power Project, 108
  Stanley Meyer, 162
  Wilhelm Mohorn, 244
Maxwell, James Clerk, 130
McAlister, Roy, American Hydrogen Association, 167
McKie, Scott, electric-power resonance, 50–52
Mead, Franklin B. Jr., electromagnetic radiation energy, 131
Media Action Grassroots Network, The, 199
media and debunkers, 189–199, 203, 205–207, 208–209
Meier, Martin, 247
methane, sources of, 37
Meyer, Stanley, water fuel cell, 161, 162–163, 164
Michelson, Albert and Edward Morley, ether drift, 129–130
Michrowski, Andrew, 161
Miller, Carmen, 14

Miller, Dayton, ether and the Earth's motion, 131
Mills, Randell 148, 151–152
  BlackLight process, 149, 154–157
  plasma-to-electricity, 150
Milton, Richard, *Alternative Science: Challenging the Myths of the Scientific Establishment*, 199
Mohorn, Wilhelm, dehydration units, 243–247
Mohorn's Self-running Wall Dryer, 243–247
Moray, John, 67, 68–69
Moray, T. Henry, 65
  Moray valve, 66–67
  Radiant Energy Device, 68, 69–70
Moray, T. Henry and John Moray, *The Sea of Energy in Which the Earth Floats*, 67
Morley, Edward and Albert Michelson, ether drift, 129–130
Mount, Andrew, 1, 3, 226
Muller, Bill, 13
  Magnet Man, 4–8
  prototype, 171
  research, 9–11
Muller, Hartmut, 72
MythBusters, 190–192. *see also* media and debunkers

N

Nagasaki, 12
Nanosolar, 177–178
nanotechnology, 177–178
NASA (National Aeronautics and Space Administration), 156
National Aeronautics and Space Administration (NASA), 156
nature
  laws of, viii
  theories of, viii
naysayers. *see* media and debunkers
NEC (New Energy Congress, The), 173, 177, 180, 229
negentropy, 98
NEM (New Energy Movement), 18
NEMCA (New Energy Movement Canada), 18
New Energy, 1, 22–26, ix
New Energy Congress, The (NEC), 173, 177, 180, 229
New Energy Movement Canada (NEMCA), 18
New Energy Movement (NEM),
New Energy Movement (NEM), 18–19, 49, 232. *see also* energy, new technology
  explosion of solutions, 229
  I Choose Love, 234
  problem is political, 43
  as a social movement, 228
  Terry Sisson, 216
New Heretic, The, Eugene Mallove, 197–199
nuclear reactions, chemically-assisted (CANR), 140
nuclear reactions, low energy (LENR), 140

O

O'Leary, Brian, 16, 195, 197, 228
open sourcing, 218
  benefit to inventors, 224
  global ripple effect, 225
  simultaneous Internet and press conferences, 221
overfishing, ix

P

Pacheco, Francisco
  and *60 Minutes*, 193
  hydrogen from water generator, 192–193
"Parallel Path Magnetic Technology for High Efficiency Power Generators and Motor Drives", Joseph Flynn, 173
patents, 218, 222–223
photovoltaics, solar, 13
physics
  laws of, ix
  plasma, 181
Pinto, Fabrizio, revolutionary energy production, 132
Planetary Association for Clean Energy, 195
PlasmaArcFlow, 167

Pons, Stanley and Martin Fleischmann, cold fusion, 141–142
power
  and radioactive byproducts, 34, 37
  solar, 35. *see also* space, solar energy
Progressive Engineer, 202
Puharich, Andrija "Henry", splitting water molecules, 161, 225

Q

quantum gravity. *see* zero-point energy
quantum vacuum, 170
quantum-leap energy, 24–25
Quasiturbine, 203–204

R

Rance River hydro project, 110, 113
Rauen, Ken, 119, 176
  heat engine, 145
Rauscher, Elizabeth, 226
Reclaim the Media, 199
Reich, Wilhelm, orgone accumulator, 88–90
resonance, 50
  and new energy science, 126
  vibratory frequency, 124–125
Resources, x
resources, natural
  stakeholders, 219–220
  universal abundance of energy, 49, 227
revolution
  clean energy, 227
  quantum-leap, 25–41, 48–49, 221
Rhodes, William, hydrogen/oxygen gas, 167
Robey, James, 159, 168
Roosevelt, Franklin D. president of United States, 87, 108
Roschin, Vladimir and Sergi Godin, 174–175
Rosenthal, Walter, 98
Russell, Walter, 82
  Russell Optical Dynamo Generator, 87
  The Secret of Life, 88
  universal electric energy, 86
  The Universal One, 88

S

Saint-Hilaire, Gilles, photo-detonation, 203
Santilli, Ruggero, magnetically-polarized gases, 167
Schaffranke, Rolf, 12–13
  absence of ether wind, 130
  Ether Technology, 11
Schauberger, Viktor, 34, 72
  antigravity, 74–75
  biotechnical machines, 76
  centripetalflow turbine design, 79–80
  implosion-based energy generator, 77–78
  inward-spiraling vortex motion, 73–74
Scheer, Hermann, *Energy Autonomy*, 33–35
Sea of Energy in Which the Earth Floats, The, T. Henry Moray and John Moray, 67
Searl, John R. R., 173–174, 175
Secret of Life, The, Walter Russell, 88
Shape Power, Dan A. Davidson, 50
Shin Deok-Ho, 226
Shoulders, Ken, electrical output generating systems, 183
Sigma, Rho, *Ether Technology*. *see* Schaffranke, Rolf
Silverthorne, Andrea, Kennedy and the Passamaquody Tidal Power Project, 109–113
Sines, Ed, 179
Sisson, Terry, MadeByKids Foundation, 216
Smalley, Richard, 213
sonofusion. *see* Stringham, Roger
space, 24–25, 49, 170
  solar energy, 13, 35, 128–129, 177–178, 185. *see also* energy, new technology
SPAWAR (US Space and Naval Warfare Sysems Center), low energy nuclear reactions, 145
SRI International, 145
StarTech, ionized air, 185
Steorn, 170
Sterling, Bruce, smog sensors, 231
Stirling engine, 185

Stolper, Thomas E.
  Genius Inventor, 148
  Randell Mills biographer, 155
Stringham, Roger, 134
  cold fusion, 139–140
  safe-fusion approaches, 145
  transient cavitation bubbles, 135–137
Swartz, Mitchell
  cold fusion, 144
  nuclear reactions tritium-free, 146
Sweet, Floyd "Sparky"
  magnetic energy generator, 93–94
  Vacuum Triode Amplifier, 95–100

T

Taleyarkan, Rusi, bubble-fusion, 138
Teal, Bob, 105
technology
  alternative. *see* energy, new technology
  current fossil-fuel based, 26–27
  globally shared, 221
  new. *see* energy, new technology
Tesla, Nikola, 197
  alternating electrical current, 56–57, 58–59
  breakthrough discoveries, 54
  free energy, 59–60, 63
  lighting for 1893 World's Fair, 59
  power of electromagnetic waves, 61–62
The Structure of Scientific Revolutions, Thomas S. Kuhn, 148
Thermal Diode for Energy Conversion, 179
thermodynamics
  first law of, 170
  laws of, viii
thermoelectric, 178
Thornton, Mike, 176
toxins, polluting, ix
tritium, 146
Trombly, Adam, 17
truck, water-fueled, ix
Truman Memoirs, The, Adm.William D. Leahy, 12

U

United States, faces unprecedented challenges, 238
Universal One, The, Walter Russell, 88
US Department of Defense, 208
US Department of Energy, 179
US Space and Naval Warfare Sysems Center (SPAWAR), low energy nuclear reactions, 145

V

Valone, Thomas, 119, 228
  abundant experiments, 131
  and Russian scientists, 175
vehicle, electric, ix, xi
Vesperman, Gary, 159
von Braun, Wernher, 11

W

Walter, Carl W., 152
Water as Fuel, Sir Anthony Griffin, 164
water-fueled systems, prototypes for, 160–168
Watson, Michael, 100
Westinghouse, George, 58
wind turbines, 180
Wolff, Milo, Space Resonance theory, 124
Wong, John, open sourcing, 218
World Environment Day, 177

Y

Yamani, Sheikh, 17

Z

zero-point energy, 15, 52, 132, 170, viii
Zinsser, Rudolf G., 10